Cathedral Mountain in Yoho National Park.

Credits

Design and maps: Dawn Huck

Contributing authors: Heidi Henderson and Philip Torrens

Contributing editors: Peter St. John, Doug Whiteway

Editorial assistance: Roy Carlson, Catherine Carlson, Ben Gadd, Brian Hayden, Grant Keddie and Ken Klein

Contributing photographers: Dennis Fast, John Harvey, Jerry Kautz, Jack Most, Peter St. John, Allan Taylor and Ian Ward.

Contributing artists: Amanda Dow, Barbara Endres, Linda Fairfield, Leah Pipe

Prepress and printing: Friesens, Canada

Front cover photographs: Spring in the Western Rockies along the Vermilion River in Kootenay National Park, and an elasmosaur skull at the Courtenay Museum. Both by Dennis Fast.

Back cover photographs: Below: Long Beach, by Jack Most; above, Western Rockies, by Peter St. John
Marpole bone harpoon on spine, by Roy Carlson

Opening page: Cathedral Mountain, by Dennis Fast

Library and Archives Canada Cataloguing in Publication

Huck, Barbara
    In search of ancient British Columbia / Barbara Huck.

Includes bibliographical references.
Contents: v. 1. Southern B.C.
ISBN 1-896150-05-5

1. Paleontology--British Columbia--Guidebooks.
2. Fossils--British Columbia--Guidebooks. 3. Geology--British Columbia--Guidebooks. 4. British Columbia--Antiquities--Guidebooks. 5. British Columbia--Guidebooks.
I. Title.

FC3807.H83 2006   971.1'01   C2006-906239-0

The partners at Heartland Associates wish to express our gratitude to Manitoba Culture, Heritage and Tourism for its continued assistance and support.

Heartland Associates Inc.
PO Box 103 RPO Corydon
Winnipeg, MB R3M 3S3
hrtland@mts.net   www.hrtlandbooks.com
5 4 3

Mixed Sources
Cert no. SW-COC-001271
© 1996 FSC

FSC

ENVIRONMENTAL BENEFITS STATEMENT

Heartland Associates Inc saved the following resources by printing the pages of this book on chlorine free paper made with 10% post-consumer waste.

TREES	WATER	SOLID WASTE	GREENHOUSE GASES
6	2,871	174	596
FULLY GROWN	GALLONS	POUNDS	POUNDS

Calculations based on research by Environmental Defense and the Paper Task Force. Manufactured at Friesens Corporation

In Search of
Ancient British Columbia

Volume 1: Southern B.C.

By Barbara Huck

with Heidi Henderson and Philip Torrens

Heartland
Winnipeg, Manitoba

Printed in Manitoba, Canada

Acknowledgements

WRITING THIS BOOK has been like birthing an elephant; or perhaps, apropos of its content, like birthing a Columbian mammoth. It has taken more months—indeed more years—than I like to contemplate, as well as the time and talents of literally dozens of people.

It's hard to know where to start, but perhaps it makes sense to begin with Roy Carlson, now professor emeritus at Simon Fraser University and widely considered to be the dean of British Columbia archaeologists. From the day of our first meeting at SFU, he was supportive and enthusiastic, and ensured that the project was truly launched by loading me down with dozens of publications published by SFU and the University of British Columbia—including such classics as *Early Human Occupation of British Columbia*, which he edited with Luke Dalla Bona. For readers interested in delving further into the fascinating history of this complex province, this and many others excellent books are listed in the bibliography at the end of this volume.

When at last the beast was about to be birthed, Dr. Carlson was also good enough to review several chapters (setting me straight over the details on several sites, adding information on others and making comments, which led to what I hope are positive changes). He was also willing to share a number of his excellent images; I am deeply grateful for all this assistance.

I appreciate, as well, the time many others spent reviewing sections of the manuscript; these included archaeologists Brian Hayden and Catherine Carlson of SFU, and Grant Keddie, curator of archaeology at the Royal BC Museum; as well as geologists Ken Klein of Thompson River University and Ben Gadd, author, educator and Rocky Mountain expert.

I am indebted, as well, to many people who took the time to meet with us, guide us to specific sites and answer our questions during our many trips around the province, and later, to review drafts of the site texts. These included Deborah Griffiths and Pat Trask of the Courtney Palaeontological Museum; Graham Beard, founder of the Vancouver Island Paleontological Museum; Elida Peers, executive director of the Sooke Museum and Archives; Helen Oldershaw of the Friends of Beacon Hill Park; Nancy Wong and Egan Davis of the VanDusen Botanical Garden; Linnea Battel, curator of the Xá:ytem Longhouse Interpretive Centre; Dave Langevin, co-founder of the McAbee Fossil Site; Margaret Holm and Brenda Baptiste of the Nk'Mip Desert Cultural Centre and Joanne Muirhead, executive director the Osoyoos Desert Society.

Though I do hope people will eventually sit down and actually read this book, perhaps more important—at least initially—to the enjoyment of it are its images and its magnificent design. These took the efforts of literally dozens of very talented photographers and artists, as well as the magical touch of our designer, Dawn Huck.

Our photographers included Dennis Fast, whose wonderful images grace the front cover as well as many of the inside pages; Ian Ward, who climbed mountains and scaled valleys to get dozens of remarkable off-the-beaten-track photos; Jerry Kautz, who with his wife Violet scaled Mount Wapta to photograph the Burgess Shale; Jack Most, who made several lengthy trips with Nancy, his wife (and my sister), to such far-flung places as Yuquot, Mount Myra and the Horne Lake Caves. I am also grateful to Allan Taylor, who allowed us to use his incomparable photographs of many sites on the Lower Mainland, and John Harvey, whose closeups of marine life and waterways taken during his prodigious kayaking trips allow a view of the province's coastal regions that would otherwise be missing. With an artist's touch, Dr.

David Blevins captured the many moods of Burns Bogs, which we were delighted to include.

Thanks too, to Barbara Winter, curator of the Museum of Archaeology and Ethnology at Simon Fraser University, who photographed a number of artifacts for us, and Kerry Lange, who allowed us to use a rare and beautiful photo of the Beacon Hill camus fields; to Holly Lenk of Tourism Victoria, who made the winsome photograph of Victoria Harbour available and to Graeme Webster of Vancouver, who helped us track down and photograph the Beach Grove Site. The *Victoria Times-Colonist* allowed us to use Darren Stone's remarkable photo of blue herons; John Leahy shared a number of his exquisite images of the fifty-million year old fossils at the McAbee Site and Marja-Leene Rathje allowed us to use her lovely photographs of Hornby Island.

In Search of Ancient British Columbia is also greatly enhanced by its artwork. Though in some places it has hardly changed, much of the province is very different than it was even a century ago. To aid in imagining the past, we used archival paintings, mainly from the National Archives in Ottawa, and modern interpretive art. The latter, which greatly assists in imagining the life and the landscape as it once was, employed the talents of several artists we are privileged to work with on a regular basis. Acclaimed artist Gordon Miller allowed us to use his painting of Crescent Beach, which I have long admired. Archaeologist and artist Amanda Dow created a series of superb watercolors that illustrate aspects of culture and society far more clearly than words alone. Artist Linda Fairfield painted and sketched environments and species long extinct. And archivist and artist Leah Pipe created symbols of the art and culture found in each region. Explanations of those symbols can be found on page 6.

But marrying all these elements into a whole took the vision, inspiration and talent of our designer, Dawn Huck. The result is not only beautiful, but enormously compelling, a book that I hope will open doors to dimensions of British Columbia that have been too often overlooked in the past. Dawn also created the dozens of lovely maps that appear throughout.

Finally, I want to thank Philip Torrens and Heidi Henderson, respectively writer and paleontologist, who took on several of the most challenging sites in the book and produced wonderfully informative and readable entries. Working with them was a pleasure and they provided a great lift at a time when the project seemed all but overwhelming.

I am also grateful to Lois Howard, our office manager, who took on endless additional work as I spent months mired under stacks of books and papers and journals; to one of Canada's great editors (and writers), Doug Whiteway, who went over the book with an eye for all the little slips that so easily evade one; and last but never least, to my husband and best friend, Peter St. John, who accompanied me on dozens of trips through B.C.; took hundreds of superb photographs; read draft after draft; created the glossary; discussed the book *ad nauseum*, while somehow summoning endless enthusiasm.

I thank you all, and also those I may inadvertently have overlooked. Without all your help, this book would have been years more in gestation.

Barbara Huck

Leah Pipe's Drawings *featured at the beginning of every chapter*

The Kootenays: Spear decorations such as these were often made with the tail feathers of magnificent bald eagles. A similar (though more modern) ornament at Canada's Museum of Civilization is made of a long piece of colored cotton with twenty-two eagle feathers attached to one edge. The Ktunaxa hunted deer, elk and bison with spears beginning more than 11,000 years ago, and though their weapons changed over the millennia to atlatls and then bows and arrows, they continued to use spears for fishing until the arrival of Europeans. The Ktunaxa still use this symbol on their website today.

The Thompson Okanagan: This pictograph of a turtle, which in many cultures represents the Earth, symbolizes a legend among the Syilx or Okanagan people of Nk'Mip. They have, in fact, chosen this symbol, on a spear point, for the attractive logo on their award-winning wines. Pictographs, which speak of the long heritage of the Nlaka'pamux, Secwepemc, Shuswap and Okanagan people throughout the Interior Plateau.

The Fraser Canyon: Beads of bone and dentalia shell speak volumes about the people of the middle Fraser River. Time consuming to manufacture (in the case of bone) or created of exotic materials (as in dentalia shell), they tell of wealthy, sophisticated cultures with the leisure, the status or the trading expertise to indulge in such ornamentation. Keatley Creek was such a society. Bone beads and shell buttons, used to decorate garments or blankets, were also found in the wealthy and stratified Marpole societies of the lower Fraser River.

The Fraser Valley: One of the things that most set the sophisticated societies along the Fraser apart from others was their art. This beautifully rendered frog is an example of the both mastery of the artist and the awareness, even centuries ago, of frogs as environmental bellwethers.

The Fraser Delta: Beautifully carved of wood or created of soapstone, as this magnificent eagle bowl was, zoomorphic bowls represent a pinnacle of form and function. Stone bowls, in particular, often combined sophisticated art with a ceremonial purpose. Leah's lovely rendering of this bowl, as well as the spindle whorl and the whale ornament that preface the Gulf Islands and Vancouver Island chapters, were inspired by the work of B.C. artist Hilary Stewart, whose drawings have brought the beauty of the past within reach of today's readers. Some of her many books are listed in the bibliography on page 304.

The Gulf Islands: As shown in the painting by Paul Kane on page 139, the creation of magnificent blankets was an art practiced for centuries, and perhaps millennia, by Salish women on all sides of the Strait of Georgia. Spindle whorls were used to assist in twisting or spinning the wool, which was made from mountain goats, wool dogs and plants. Most spindle whorls were apparently made of carved hardwood, but since these did not last long once they were discarded, it is whorls of bone, such as this one (which was carved on both sides) that have usually been found.

Vancouver Island: This handsome whale is an ornament, appropriately carved of whalebone. With a hole in its upper fin, it was clearly meant to be worn. Though this particular carving was excavated on the northwest coast, ornaments, carved bowls, pestles, charms and combs of a similar style have also been found on Vancouver Island.

Table of Contents

A note for our readers: We struggled for some time over the geographical layout of the book. We finally decided that while the majority of the population undoubtedly lives in the province's southwest, geologically, paleontologically and even—given the data available today—archaeologically, eastern B.C. long predated the west. To be true to our subjects, we decided that In Search of Ancient British Columbia had to do the same.

Building British Columbia

MANY WHO HAVE VISITED British Columbia (and indeed, quite a number who live there) have described the province as "otherworldly". Citing its towering mountains, wild Pacific shores and magnificent old-growth forests, its arid Interior Plateau, great rivers and unmatched salmon runs, they call it a place apart from the rest of Canada. And, geologically speaking, they're right. With the exception of the province's eastern edge, British Columbia is a place apart. In fact, it's many places apart.

It should come as no surprise, then, that building British Columbia was lengthy and complex. Even the geological descriptions of it are lengthy and complex, having to do with plate tectonics, accretionary complexes, magmatic arcs and crustal stretching. Geologists should not be blamed for their complicated jargon, however. Piecing together the province's formation has been a convoluted process that has taken the past century and more, and involved the work of many individuals. Even today, there are still areas of B.C.'s geological map that have question marks on them. Fortunately, the mountains, valleys, deserts and seashores have provided portals that, as our tools increase in sophistication, have increasingly assisted in piecing together much of the ancient past, not only of British Columbia, but of our planet.

One of the best known of these windows on time is the Burgess Shale (see page 42), an ancient undersea archive that today is hundreds of kilometres from any ocean and 2,300 metres above sea level. Reached via a demanding alpine trail in Yoho National Park in the Canadian Rockies, this 513-million-year-old site offers an almost perfectly preserved picture of what may have been the Earth's most important eruption of life.

Melanie Froese

The Cambrian Explosion, which according to zircon dating began 542 million years ago, was an orgy of evolution. Yet, in geological terms, it was equal to the blink of an eye. In a period of about five million years, all the major body plans of animals evolved, including the ancestors of vertebrates or animals with backbones. Everything since, the late biologist Stephen Jay Gould once wrote, has been nothing but variations on a theme.

For decades, the Burgess Shale appeared to stand alone, the only evidence of this remarkable evolutionary eruption. Even more disconcerting, there seemed to be no introduction to this revolutionary period. At least twenty million years seemed to be completely missing from the paleontological record. Well, no more. Scientists now know that "Ediacara fauna", transitional creatures named for the red rock in southern Australia

Trilobites, which may be the world's favorite fossils, emerged with the Cambrian period and long outlasted it.

where they were found, as well as worms and even creatures with elemental backbones found in China and Namibia, inhabited what is now known as the Ediacaran period, between 600 and 543 million years ago. Slowly, the pieces are being filled in.

Across the Kicking Horse Valley

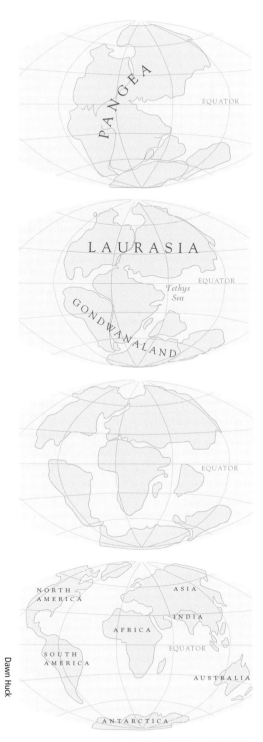

PANGEA

EQUATOR

LAURASIA

Tethys Sea

EQUATOR

GONDWANALAND

EQUATOR

NORTH
AMERICA

ASIA

INDIA

AFRICA

EQUATOR

SOUTH
AMERICA

AUSTRALIA

ANTARCTICA

Pangea likely looked like the top map about 225 mya;
the Atlantic Ocean was created as the supercontinent
broke up between 200 mya (Map 2) and sixty-five mya
(Map 3). The bottom map shows the continents today.

Dawn Huck

from the Burgess Shale, the Mount Stephen
Trilobite Beds (see page 47) tell more of the
Cambrian story. Like their counterparts on
Fossil Ridge, the trilobite beds are also high
in the Rockies today, but at one time lay along
the shifting shorelines of embryonic North
America. A half-billion years ago, these shallows
brimmed with life, in sharp contrast to the
barren landscape of the continent. Without
trees or grass or plants of any kind to anchor
the soil, the wind and rain wore at the bones
of the earth. Gushing rivers carried sand and
silt from ancient mountains in the continental
core to the sea. Layer upon layer, the coastal
sediments grew, littered with the distinctive
shells of dozens of trilobite species, as well
as the many other wonderful creatures that
evolved over time.

Some of these creatures can be found atop
another mountain in the Kootenays, at Top-of-
the-World Provincial Park (see page 76). Here,
along with cephalopods and nautili—distant
ancestors of the chambered nautilus that can
still be found in the Pacific and Indian Oceans
—were receptaculitids, large balloon-like plants
that were once believed to be animals. Found
together, these fossils allow paleontologists to
date the rock to the Ordovician period, between
about 505 and 440 million years ago. In fact,
simply by studying trilobites, geologists can
almost invariably date any period from the
early Cambrian about 540 million years ago
(mya) to what has been called "the granddaddy
of all mass extinctions", the still largely unex-
plained cataclysmic event that ended the
Permian period 248 million years ago. But
we're getting ahead of ourselves.

By about 280 million years ago, sliding on
their tectonic plates, the continents were again
coming together to form a huge supercontinent
we know as Pangea, "all lands". This was not the
first congregation of continents in the Earth's
long history, far from it. Geologists now believe
that at least four other supercontinents had
previously formed: Kenorland (which existed
between about 2.45 and 2.1 billion years ago);

Columbia (~1.8-1.5 billion years ago); Rodinia (~1,100-700 million years ago) and the short-lived Pannotia, which existed between about 600 and 540 mya. Because much of the evidence of older supercontinents is destroyed each time more recent ones are formed, we know far more

Dawn Huck

Comparing the similarities and evolutionary differences among species has aided paleobiologists in reconstructing the constituent parts of Pangea.

about Pangea than we do about any of the others.

We know that Laurentia (or ancient North America) occupied the northwest quadrant of Pangea (see map series on page 11) which reclined on its side, compared to its modern orientation, touching the equator. The land that would one day be southeastern British Columbia lay just off-shore, facing northwest, beneath a shallow continental sea. And what would eventually be the rest of the province was either somewhere out in the Pacific, as embryonic island arcs or terranes far to the southwest, perhaps where Japan is today, or—as in the case of what would one day be southwestern Vancouver Island—lay deep within the Earth's molten core, waiting to be born as undersea volcanic ridges where mid-ocean tectonic plates meet. The Pacific Ocean covered most of the Earth's surface and the Atlantic was yet to be born in any meaningful way.

The Permian extinction, which scientists believe destroyed nearly ninety-five per cent of the Earth's creatures, opened the door to new forms of life. These included ammonites—the name comes from the uncanny resemblance of the fossils of these ancient sea creatures to coiled rams' horns, the symbol of the Greek god Ammon—which were just beginning a global ascendancy. The fact that different sub-species were found almost everywhere was of particular importance to geologists, for they could match fossils found in one place with those found in another. In the case of British Columbia, it was ammonites—specifically Jurassic ammonites—that allowed scientists to piece together a map of the world 200 million years ago. Matching the ammonites found in abundance in the rocks of South Moresby Island on Haida Gwaii with other similar creatures on the other side of the Pacific, led to the realization that B.C.'s coastal islands, including Vancouver Island, were once located on the other side of the planet.

In fact, thanks to the work of Canadian geologists Jim Monger and Tim Tozer, along with their American colleague Charlie Ross, it is now clear that much of British Columbia was once a series of volcanic islands born far to the south in the primeval Pacific. Beginning about 270 million years ago, these islands began to co-alesce into what the geologists term "terranes"—island arcs rather like the Phillipines today. Gliding slowly northeast, some of them merged into a small continent, which by 200

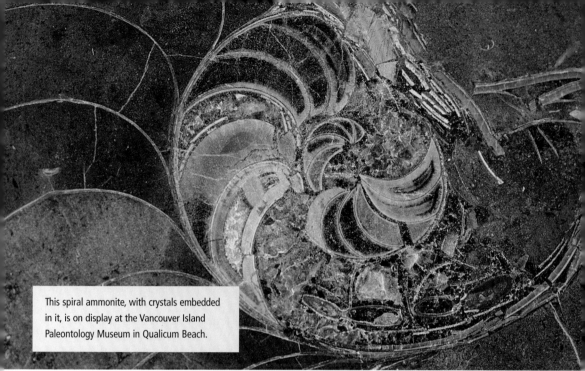

This spiral ammonite, with crystals embedded in it, is on display at the Vancouver Island Paleontology Museum in Qualicum Beach.

Dennis Fast

million years ago lay just west of the giant supercontinent Pangea.

This was the Jurassic, the beginning of the long ascendency of the dinosaurs. Because so much of British Columbia was underwater, it was believed until very recently that the province had a dearth of dinosaurs. So no one bothered to look for them. No one, that is, except the believers—who were all, it turns out, under the age of twelve. Mark Turner and Daniel Helm were among them. The boys were tubing on Flatbed Creek near Tumbler Ridge in the province's northeast when, having been spilled into the water, they spied what they were certain were dinosaur tracks in the rock along the river. But when they told their families, the discovery was initially dismissed as a childhood fantasy. Even when induced to look at the tracks, the skeptical adults were unconvinced. It took a paleontologist to make everyone realize that the boys had, indeed, found the tracks of an ankylosaur, an armored dinosaur that turned out to be the first of many species found in the region.

Even more surprising, the Tumbler Ridge dinosaurs were discovered to be about ninety-five million years old, predating their Alberta cousins by between ten and thirty million years. The result has been dinosaur fever in Tumbler Ridge, which will be investigated in Volume II of *In Search of Ancient British Columbia.*

However, as we will see in this volume, Daniel and Mark were not the first youngsters to have found an ancient species in British Columbia. Courtenay's Heather Trask was only eleven when she and her father Mike took a late autumn walk along the Puntledge River in 1988. There, in the shale along the river, they discovered the neck bones of what turned out to be an elasmosaur, a huge marine reptile (see page 271). Since then, mosasaurs have also been found in the region and Courtenay is now home to an excellent paleontology centre.

As dinosaurs and plesiosaurs flowered worldwide, Pangea began to come apart at the seams. Laurentia, ancient North America, slid northwest, creating the Atlantic Divide and colliding, between 185 and 170 million years ago, with the smaller continent that was riding northeast on the Pacific Plate. This smaller continent had begun as three "terranes" —the Quesnel, Cache Creek and Slide Mountain— and grown by picking up others. As indicated earlier, they had joined together and now slammed obliquely onto the edge of ancient North America, initiating the continent's second era of mountain building. (The first, which had taken place more than 1.7 billion years ago, created an arcing chain of lofty mountains in the heart of the North American continent. Over deep time, these would wear down to become the Precambrian or Canadian Shield.)

Like a wedge, the edge of Laurentia peeled the lighter rock of the island terrane off the underlying oceanic plate, cementing it onto the continental margin. Today, this broad piece of territory is known as the Intermontane Belt or the Interior Plateau, while the continental zone of impact, which was compressed, thickened and, in places, heated to the melting point, became the Omineca Belt of east-central BC.

Coastal collisions like this continued over millions of years and inland, the new territory served as an enormous bulldozer. Rumpling the sedimentary layers like tinfoil, the Columbia Mountains were created, while farther east, the pressure caused the land to buckle and jackknife skyward as the main ranges of the Rockies were created about 140 million years ago.

But the British Columbia we know today was not yet complete. For a time, the Pacific Plate slowed and began to move almost directly north. The North American coast was also moving northward, but more quickly, and like a truck passing a car on the inside lane, this caused the Coast Belt—a wide band of territory that had not yet firmly attached itself to the continental edge—to slide south relative to the huge continent. This brought a belt of smaller terranes into contact with the continent's western edge. (Parts of these terranes would eventually move north again, but bits and pieces would remain just west of the Fraser Canyon (see pages 141).

Then as the Pacific Plate shifted course once more and began to move northeast again, the North American plate switched to a southwestern track. The result, between eighty-five and ninety million years ago, was a direct collision between North America's west coast and Wrangellia, a huge terrane that includes today's Vancouver Island and Haida Gwaii.

The Rocky Mountain Trench, seen here near Fairmont Hot Springs, makes the concept of crustal stretching easy to understand.

Inland, the melted crust of the collision zone created the upheaval of the largest mass of granite in the world along the Coast Belt, and farther east, caused the front ranges and foothills of the Rockies to rise.

Finally, perhaps sixty million years ago, all the major elements were in place. But there were still adjustments to be made: millions of years of volcanic activity and subsequent erosion in the Okanagan Highlands (see page 112); crustal stretching that would create the Rocky Mountain Trench (see page 48), and the birth of a line of young volcanoes in the Cascade arc, many of which are still volatile.

By two million years ago, beautiful British Columbia had its great forests, interior grasslands and deserts and its remarkable diversity of land and sea life. Now it needed only its finishing touches, as well as its unique cultures that would set it apart from everywhere else on Earth.

The finishing touches came first, courtesy of the Pleistocene, Earth's long period of global cooling that would impact everything in British Columbia from its mountain tops to its valley floors, and provide the causeway over which many species of animals and, eventually, humans would populate a new world.

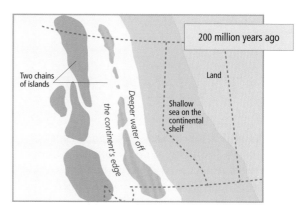

200 million years ago

Two chains of islands

Land

the continent's edge

Deeper water off

Shallow sea on the continental shelf

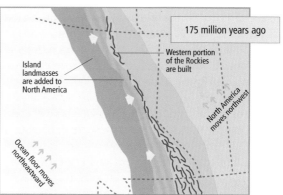

175 million years ago

Island landmasses are added to North America

Western portion of the Rockies are built

North America moves northwest

Ocean floor moves northeastward

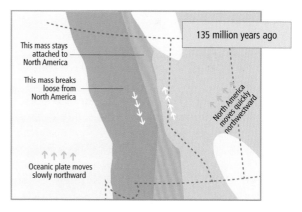

135 million years ago

This mass stays attached to North America

This mass breaks loose from North America

North America moves quickly northwestward

Oceanic plate moves slowly northward

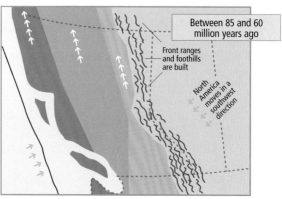

Between 85 and 60 million years ago

Front ranges and foothills are built

North America moves in a southwest direction

As these illustrations show, British Columbia was built over a period of nearly 150 million years. As a result of a complex series of movements by tectonic plates, island arcs and terranes were accreted or added to the western edge of North America.

Dawn Huck, after maps by Ben Gadd in his *Handbook of the Canadian Rockies* (second ed.)

The Icing on British Columbia

MENTION BRITISH COLUMBIA and visions of green fairways at Christmas and February daffodils come to mind. Yet B.C. has more permanent ice, and more glaciers, than anywhere south of Ellesmere Island. And while it seems clear the province's mountainous terrain is responsible for this apparent dichotomy, scientists from a variety of disciplines have discovered that British Columbia is involved in other icy contradictions. These anomalies are making them rethink their interpretations about Earth's most recent lengthy glacial incursion, which many people call "the ice age".

But first, a short lesson on glaciation. We tend to think of the temperatures we humans have known for the past 2,000 or 3,000 years as normal. Even in British Columbia—with the possible exception of the southwestern coast—that means warm summers and comparatively cold winters. We grow concerned when global mean temperatures rise between .6°C and 1.1°C, as they have since the Industrial Revolution. But for most of Earth's long history, global temperatures were between six and eight degrees Celsius warmer than they are today, according to University of British Columbia geographer Michael Pidwirny.

That kind of global warming might not seem surprising when one imagines the Jurassic or early Cretaceous periods, which we picture with dinosaurs in a subtropical setting, for the Earth's continents were clustered around the equator. But even fifty million years ago, when British Columbia sat at approximately the latitude it does today, rock solid evidence in the Okanagan Highlands (see page 112) makes it clear that British Columbia was still subtropical, while in the Canadian Arctic, trees with trunks a metre in diameter grew.

This is not to say, however, that the global climate has always been warm. Quite the contrary. In the past billion years, Pidwirny writes, "glacial periods started at roughly 925, 800, 680, 450, 330 and 2 million years before present …". The most severe period of global cold, he believes, was about 800 million years ago, when glaciers came to within five degrees (or 550 kilometres) of the equator.

It is the most recent of these glacial ages—the Pleistocene epoch, which began about two million years ago —that we know best,

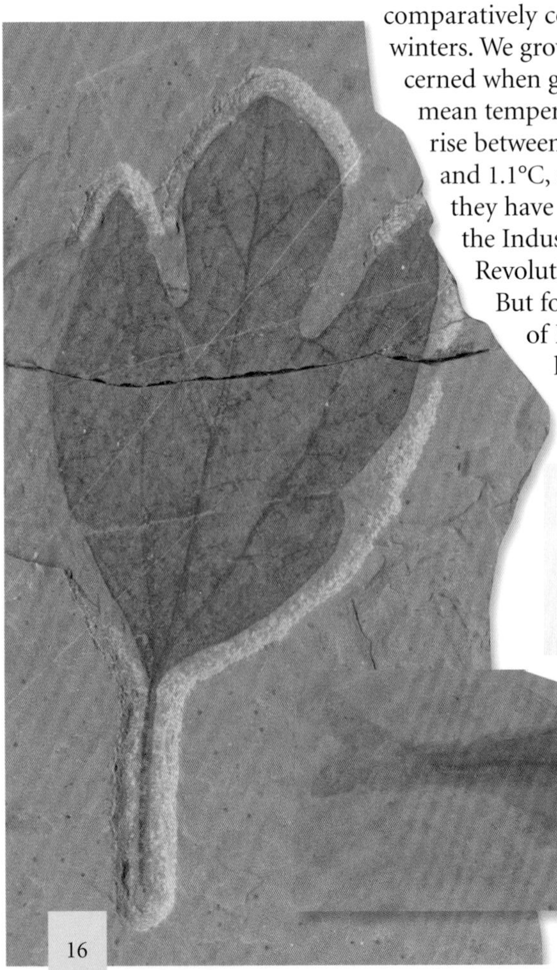

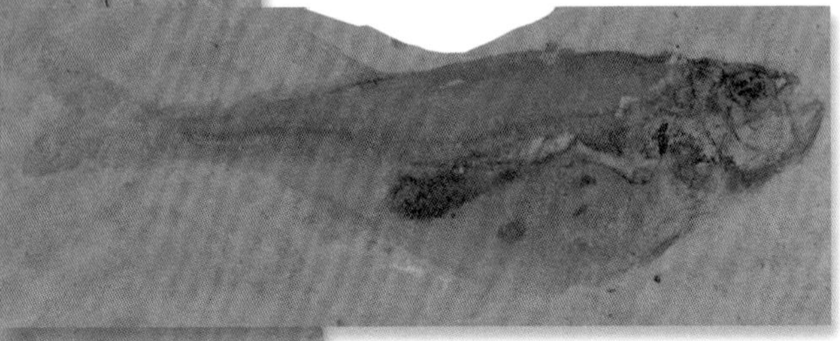

Fifty-million-year-old fossils from the Okanagan Highlands, including trilobed sassafrass leaves and mooneye shown here, show that the Earth was once warmer than it is today.

Photos: John Leahy

Ice Age Climate Variability

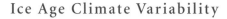

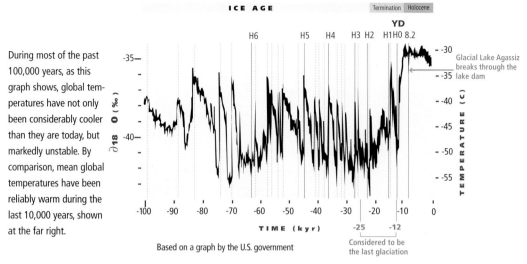

During most of the past 100,000 years, as this graph shows, global temperatures have not only been considerably cooler than they are today, but markedly unstable. By comparison, mean global temperatures have been reliably warm during the last 10,000 years, shown at the far right.

Based on a graph by the U.S. government

Considered to be the last glaciation

of course. Yet even during our current "ice age", as the interlude of global warmth over the past 10,000 years makes clear, the global climate has not been consistently cold. It has simply been cold more often and for periods of longer duration than it has been warm.

Climatologists believe that our current glacial age has involved between seventeen and twenty glacial incursions and that we are living today in an interglacial period that is very likely—notwithstanding the current concerns about global warming—nearing an end. In fact, some climatologists believe that a spike in global temperatures may be precisely what will lead to the next big chill.

Yet if the first discernable glacial age began less than a billion years ago, and geographers have been able to count six glacial ages since, it's clear they are unusual in the larger context of the Earth's 4.6-billion-year existence. In other words, for more than ninety-five per cent of its history, our planet has enjoyed a warm, benevolent climate. Why, then, all this ice? And why now?

As I wrote in *In Search of Ancient*

Alberta, many theories have been proposed since the world's scientists began studying climate in the early 1800s. Currently, the most persuasive goes something like this: glacial ages are closely linked to the movement of the Earth's tectonic plates and their onset is largely governed by the patterns the continents create. As a glance at a modern globe clearly shows, the present pattern has the continents spread over the northern half of the globe in such a way that almost completely isolates the Arctic Ocean, which is centred on the North Pole. This configuration largely prevents ocean currents from reaching the most northerly regions of the planet, resulting in summers at high latitudes that can be too cold to melt all the snow and ice that accumulates each winter.

But this explanation does not account for the fact that over the past 25,000 years almost all of Canada and the midwestern U.S. were covered by huge sheets of ice for more than 12,000 years and then rapidly unglaciated. Nor does it explain the fact that, while glaciers in the Arctic and the mountains of B.C. and Alberta continue to recede, climatologists are quite certain another glaciation is on the way.

The explanations for these apparent contradictions depend on the other part of the climate equation, the astronomical part. Scientists believe that climate has a cycle of approximately 100,000 years. During a glacial age, glaciations lasting from 60,000 to 90,000 years alternate with interglacial periods of between 40,000 and 10,000 years. The 100,000-year cycle—called the Milankovitch cycle after its discoverer,

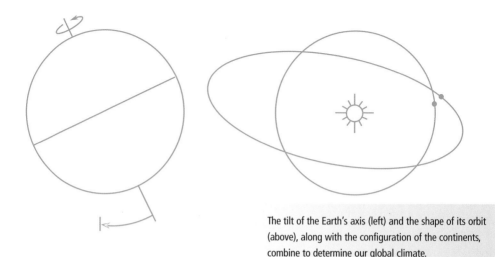

The tilt of the Earth's axis (left) and the shape of its orbit (above), along with the configuration of the continents, combine to determine our global climate.

Serbian scientist Militun Milankovitch—is determined by the Earth's orbit around the sun, which in turn is dependent on three other cycles. The first is the 105,000-year variation in the shape of the Earth's elliptical orbit; the second, the 41,000-year cycle in the tilt of the Earth's axis and the third, a 21,000-year cycle in the movement of the date at which the Earth is closest to the sun as it moves along its orbit. This last cycle, called the precession of the equinoxes, shifts forward through the months from January, through February to March and so on around to January again at the end of 21,000 years.

The cumulative effect of all these cycles for high-latitude countries like Canada is to determine the degree of variation between summer and winter temperatures. At one extreme, the northern continents will have relatively warm summers and cold winters, as they have had for the last few thousand years. At the other extreme, the same areas will experience relatively cool summers and somewhat warmer winters. It is the summer temperatures that, more than anything else other than the arrangements of the continents, accounts for the appearance and disappearance of the great sheets of ice. Cool summers that are not fully able to melt the snow accumulation from the previous winter will inexorably lead to the development of glaciers.

Glacial coring in the Arctic has recently shown that over most of the past 100,000 years Canada was considerably cooler than it is today. We must therefore go back perhaps 125,000 years, to the last major inter-glacial period, known as the Sangamon inter-

glacial, to find a similar period to the one we currently enjoy. In many ways, Sangamon British Columbia would have resembled our modern province, with forests of hemlock and spruce in the mountains and cedar on the coasts. But the mammals that inhabited this landscape were quite different than those we know today. Massive stag-moose thrived in the valleys, along with shrub-oxen, helmeted muskoxen, mastodons, mammoths and several kinds of horses. Preying on this bounty were sabre-toothed tigers, fleet and ferocious short-faced bears, and huge dire wolves. Many of these animals had distant relatives on the other side of the globe, but cut off by the surrounding oceans, they had developed in distinctive ways. Others, like *Equus*, the horse, as well as two kinds of pronghorn antelope, originated in the Americas, and had evolved over fifty million years from primitive hoofed mammals.

Mammut, the mastodon, had immigrated to North America millions of years earlier and evolved in unique ways, while the woolly mammoth had emigrated from Asia over the land bridge during the early part of the current ice age. All had one

Barbara Endres

Mastodons, distant and more primitive relatives of mammoths, were actively hunted by early North Americans.

thing in common, according to Canadian naturalist E.C. Pielou. "All these animals were descended from ancestors that had lived in North America for more than one million years in an environment devoid of ruthless, expert human hunters," she writes in *After the Ice Age*.

That innocence spelled extinction for all but the speedy pronghorn. Along with many other species, every one of these large North American mammals was to disappear forever in the two or three millennia following the most recent glaciation, which climatologists call the Wisconsinan. In their place an army of Old World species—deer and moose, black bears and grizzlies, cougars and timber wolves, all familiar with the nasty wiles of human hunters—crossed Beringia to settle in North America.

But 125,000 years ago, all this was far in the future. Life in Sangamon British Columbia was bountiful, with hot summers and cold winters (again with the likely exception of the southwest and coastal areas). And if, over the millennia, the summers grew gradually cooler and shorter than they had been, and late snows took a toll on newborn fawns, as always the fittest survived. By 75,000 years ago, however, the mountain glaciers were growing down the slopes and the high valleys were filling with winter snow that did not melt. On the west coast, both the water temperature and the sea level were dropping and icebergs could sometimes be found as far south as northern Vancouver Island. A population of walruses, perhaps cut off from their normal home in the Bering Strait by the rising land

bridge, populated the Strait of Georgia (see page 265).

Over the next 50,000 years, the glaciers waxed and waned. Then, about 25,000 years ago, the ice began to expand in earnest. By 18,000 BP, most of Canada and the northern Midwestern states were covered by huge ice sheets two kilometres or more thick. Yet in the Thompson River Valley, kokanee salmon continued to spawn (see page 108), and on Vancouver Island's Saanich Peninsula, Columbian mammoths were filling their king-sized stomachs (page 231). Clearly, parts of British Columbia remained unglaciated until long after most of Canada.

Glacial striations on some of the most southerly rock on Vancouver Island (see page 232) show that the ice did eventually cover the Victoria region, though perhaps not until 15,000 years ago. And it didn't stay long. A thousand years later, the ice sheets were already retreating north. And farther east in the Kootenays, artifacts excavated from high south-and west-facing lake terraces show that human hunters were living there between 11,500 and 12,000 BP (see page 80). In short, the last glaciation was not as simple, and perhaps not as pervasive as previously supposed, which has not only complicated the task of unravelling B.C.'s glacial history, but opened the door to other interpretations on the peopling of not just British Columbia, but North America as a whole.

The real problem is that global climate change is remarkably complex. Even during temperate periods, such as the Earth has enjoyed for the past 12,000 years, there can be spikes of global warmth and cold. And as indicated earlier, we are beginning to realize that one may well lead to the other. History has given us examples of this in the recent past. The most recent

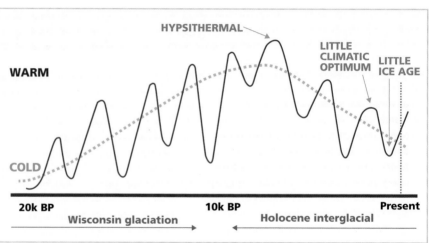

Overall, the Earth's global climate has been cooling since the Hypsithermal, at the end of the Wisconsin glaciation. However, within that trend of gradual cooling, there have been lengthy periods of warmth, such as the one we are now enjoying, as well as periods of dramatic cold.

period of global warming, which climatologists call the Medieval Warm Period or Little Climatic Optimum, varied from place to place around the Northern Hemisphere, but was evident almost everywhere between 1,200 and 700 years ago.

In British Columbia, studies of soil samples around Dog Lake in Kootenay National Park (see page 62) have shown repeated forest fires, infestations of bark beetles and regular drought conditions during this period. Far to the west, a string of fortified defensive sites grew up between 1,200 and 1,000 years ago around the southern coast of Vancouver Island (see page 236). Grant Keddie, the curator of archaeology at the Royal BC Museum, has suggested that these fortified villages "may have resulted from a more intense specialization in procuring migratory fish". But similar movements from inland sites to more bountiful riverside and coastal locations also took place in many places in the U.S., and have been attributed to drought conditions in the interior. In Britain, meanwhile, the Medieval Warm Period had definite bonuses: wine grapes were grown as far north as the Scottish border, and the Old Norse were able to settle and farm Greenland.

This balmy interval came to a rather sudden end about 650 years ago. All over the Northern Hemisphere, winters were suddenly long and cold and summers short and cool. In B.C., Dog Lake grew larger and deeper and in the forests around it species grew that thrived better in cold, damp soil. In the Coast Mountains and on Vancouver Island, glaciers that had receded to the point of being nonexistent suddenly began to form and grow. They continued to grow (with small reverses), mowing down well

established forests, until less than 100 years ago. An example is Septimus Glacier in Strathcona Provincial Park (see page 293). Studies of glacial moraines and dendrochronology (the study of tree rings) by geographers from the University of Victoria show that Septimus Glacier has lost eighty per cent of its mass since it reached its greatest extent in the early 1900s. It is therefore likely that it had disappeared entirely during the much longer Medieval Warm Period.

In Europe, the Little Ice Age meant skating parties on the Thames for Victorian Londoners, but also crop losses so severe, according to University of California anthropologist Brian Fagan, that "Alpine villagers lived on bread made from ground nutshells mixed with barley and oat flour." Finland lost almost a third of its population to starvation and disease and the 500-year-old Greenlandic colony, cut off from supply ships by pack ice around the coast, shrivelled up and died.

Today, respected climatologists are suggesting that the Medieval Warm Period and the Little Ice Age were linked and that similar conditions may be building today. The key, it seems, lies with the Gulf Stream, a "river" of warm water that flows north up the Atlantic Ocean from the Caribbean to about the latitude of Portugal, before dividing. Some of the water turns south, creating a surface current of warm water known as the subtropical gyre, while the rest continues north, creating warm winds that elevate the temperatures of Western Europe by between 5°C and 10°C. South of England, this current turns west and then, finally depleted of its warmth, sinks into the North Atlantic in two massive columns of cold water off Greenland, which serve as the main engine powering deep water currents scientists call the Great Ocean Conveyor.

However, studies by the Southhampton Oceanography Centre in the United Kingdom and the Woods Hole Oceanographic Institution in Cape Cod, Massachusetts, as well as Canadian teams of oceanographers from British Columbia, have shown two disturbing things. First, enormous rivers of fresh meltwater—some 20,000 cubic kilometres of cold, fresh water, equivalent to a layer more than three metres thick—have flowed into the North Atlantic over the past thirty years. This layer of lighter, fresh water sits on top of the ocean like a thermal blanket, threatening to slow or even stop the circulation. This may already be happening, according to Harry Bryden of the British team. His studies have shown that while one column of cold water on the Canadian side of Greenland appeared to be sinking in a normal way, the other, on the European side, is sending only half as much deep water south as previously. Further, the two columns are travelling south at different depths. And the speed of the Gulf Stream appears to have slowed by as much as thirty per cent.

Terrence Joyce, chair of the Woods Hole Physical Oceanography Department, says a continued shift in the salinity of the Atlantic and a continued slowdown in the speed of the Gulf Stream could lead to it shifting south or coming to a complete halt. And the result would almost inevitably be another prolonged period of global cooling, another Little Ice Age, or perhaps something far worse. Moreover, he believes this is not something that would take a century or more to occur. It could happen in a decade, he says. And the results could last for centuries.

The last glaciation left British Columbia with many special places, including the Sooke Potholes, below. But it also made Canada virtually uninhabitable for thousands of years.

Jack Most

Peopling the Americas

FEW MYSTERIES are more compelling than the question of just when humans first arrived in the Western Hemisphere. For decades, many archaeologists thought they had the answer. The first Americans, they believed, were skilled hunter-gatherers who travelled across the Bering land bridge in the waning millennia of the last glaciation, about 13,500 years ago. Penetrating North America through what was termed the Ice Age Corridor, a hypothetical passage between the Laurentian and Cordilleran ice sheets along the eastern foothills of Alberta that opened as the glaciers melted perhaps 12,000 years ago, the theory went, they spread rapidly through the hemisphere within about 500 years.

This idea was called the Clovis hypothesis, named for the 1932 discovery of the first of many beautifully-crafted willow leaf-shaped spear points in a pile of mammoth bones near Clovis, New Mexico. The bones, and thus the points, were later determined to be approximately 11,200 years old. As similar points were unearthed all over North America in the wake of the receding ice, and all dated to about the same period, it seemed the mystery was solved. The first Americans must have been the originators of the Clovis points, a clever, technically-sophisticated people who rapidly spread through the Americas, leaving behind their signature weapons.

But some felt the theory failed in a number of ways. How, they wanted to know, could a people accustomed for thousands of years to an existence of frigid Arctic survival so quickly adapt to the temperate and tropical conditions they would meet as they moved south through the Americas? And what, faced with the bounty of an entire hemisphere filled with animals unfamiliar with the deadly wiles of man, would prompt them to sprint through their new environment, particularly when they knew nothing of what lay ahead?

Consider, some said, that it took much later Americans, equipped with draft animals, beasts of burden, the wheel, relatively sophisticated methods of navigation and a global view, almost 250 years to cross from Massachusetts to California, a distance of perhaps 5,600 kilometres. The idea that a much less technologically advanced people could have moved from Alaska to Chile—some 16,000 kilometres of ever-changing, ever-challenging terrain and climate—in only twice as many years seemed little short of incredible.

In the past decade, the idea of an ice age corridor has gradually faded, and in its place is increasing evidence that people may have arrived in the Americas along a coastal route that skirted the southern shores of Beringia and the west coast of Alaska and British Columbia. But if this was the route to a new world, where was the evidence?

During the last two decades, a number of determined archaeologists have dedicated themselves to the search for ephemeral traces of the brief passing of an ancient people across a vast territory. Remarkably, they have found it in places as far-flung as Chile and Yukon, Brazil and New Mexico, Venezuela and Washington State. Their research has put the Clovis hypothesis to rest once and for all, but has only deepened the mystery. When did humans first arrive in the Americas? And was British Columbia their main gateway?

Melanie Froese

Clovis points led to the first theories about the peopling of the hemisphere.

The discovery, in Taima-Taima, Venezuela, of a 13,000-year-old mastodon with a stone point embedded in its pelvis, was one of many that contested the Clovis theory.

The answer to the first question — when did they arrive?—is still very much up in the air, but we do know it was earlier than 12,000 years ago. The most widely accepted body of evidence comes from a site called Monte Verde, in southern Chile. Located on the banks of a creek in wooded hills near the Pacific Ocean 800 kilometres south of Santiago, Monte Verde has convinced even the most skeptical experts that humans were living in the Americas—even some of the most remote parts of the Americas—almost 13,000 years ago, long before Clovis points were apparently invented.

Monte Verde has been described by its excavator, American archaeologist Tom Dillehay, as a camp, but it was a remarkably settled camp, more like a village. And because the site was subsequently covered by a thick layer of peat, not only bones and stones but organic matter was remarkably well preserved.

The little settlement had at least twelve rectangular, tentlike dwellings, somewhat similar to those used on the Russian plains 35,000 years ago. American and Chilean archaeologists, who began digging at Monte Verde in the mid-1970s, found crude timbers and wooden foundations still intact. Inside the huts were an array of tools, dishes, foodstuffs and medicines that paint a picture of an accomplished, proficient people who ranged widely over their home territory.

The scientists found firepits surrounded by charred wood chips, wooden lances with hardened tips and wooden bowls containing seeds. They discovered wild potatoes, a chunk of meat later identified by DNA analysis as mastodon, as well as seaweed and other plants which must have been imported from the Pacific coast, fifty kilometres away. In a wishbone-shaped structure nearby were more than twenty types of medicinal plants. But most poignant, perhaps, was the discovery of the footprint of a child in what had been soft clay near one of the firepits.

Following exhaustive scrutiny by the most critical international experts, Monte Verde was at last accepted in 1997 and its antiquity—12,800 years—verified. Though there are still some critics of the site, their criticisms are largely limited to details and small inconsistencies that Dillehay has been able to answer. Conceding the age of Monte Verde thrusts other contested sites into the limelight, including Taima-Taima, in northern Venezuela. Excavated by University of Alberta archaeologists Alan Bryan and Ruth Gruhn with the site's original archaeologist, the late José María Cruxent, it is an ancient artesian well in which a young mastodon met its death, apparently at the hands of early Americans. The bones of the animal are 13,000 years old; a stone point was embedded in the pelvic area.

In eastern South America, a much older and more controversial site is located among sandstone cliffs in the arid outback of northeastern Brazil. Pedra Furada, a huge rockshelter, has been the passion of French-born and trained archaeologist Niede Guidon for almost thirty years. And though she is universally praised for her professionalism and dedication, the results she claims have been hotly disputed. Guidon found cave paintings, crude stone tools and ash

Bering Land Bridge

Greenland Ice Sheet

Bluefish Caves, Yukon, Canada
10,000 - 23,000 years old

Cordilleran Ice Sheet

●Calgary

Laurentide Ice Sheet
ca. 13,000 years ago

The idea that early North Americans entered the continent through an Ice Age Corridor, shown in blue, is increasingly giving way to theories about coastal migration and the idea that the first wave of emigrants arrived prior to the late Wisconsin glaciation.

Clovis, New Mexico
11,500 years old

Meadowcroft Rock Shelter, Pennsylvania
12,000 - 15,000 years old

Dates are subject to debate

Taima - Taima Venezuela
13,000 years old

Pedra Furada, Brazil
30,000 - 50,000 years old

Monte Verde, Chile
12,300 - 12,800 years old

hearths which have been radio-carbon dated to more than 30,000 years.

Her critics do not contest the datings; they question instead whether the crude stone tools were in fact made by humans, or fractured by natural processes and mixed with rounded river rock by water action, and whether the fires were actually begun by humans or were simply the result of forest fires. Guidon counters that the stone tools could not have been carried to their final resting place in the area of occupation by stream action. And the charred wood is unlikely to be the result of forest fires because the cave is eighteen metres above the surrounding area. Moreover, she points out, forest fires would have been very unusual, for what is now a parched region of thorn forest and cactus was then a tropical rainforest. Here, many tree species coexisted, unlike our northern forests which are dominated by easily-ignited pines and spruce. Not only is such a rainforest less prone to fires, the charcoal Guidon found was concentrated at the rear of the cave, behind a protective wall of fallen rock.

The excavated artifacts and cave paintings seem to confirm her climate analysis. The digs have uncovered the remains of ancient tropical trees and the paintings include images of wetland animals like the capybara, a large aquatic mammal related to the guinea pig. Even the name of the national park which surrounds the Pedra Furada echoes its steamy tropical past. It's called Serra de Capivara—the Capybara Mountain Range. Despite all this, the fight for international acceptance of the site has been long and hard and, while European experts have largely accepted her findings, Americans continue to doubt the site's antiquity.

Archaeologist Richard MacNeish was not among them. MacNeish, who spent years working in Canada, headed a team from the Andover Foundation for Archaeological Research which spent three years digging at Pendejo Cave on an army firing range in southeastern New Mexico. Pendejo is a small triangular cave located at the base of a limestone cliff, 100 metres above the modern desert floor. Excavations revealed twenty-six layers of occupation, with hundreds of artifacts including primitive stone tools, charred bones of extinct animals, rock-lined fire hearths and remnants of burned campfire logs. MacNeish and his crew also found human hairs, eleven fingerprints and a horse toe bone with a bone projectile point embedded in it.

MacNeish also worked in Mexico and South America. In southern Peru, high in the Andes foothills, he found three major occupation layers at a place called Pikimachay Cave. In the middle layer, bone projectile points and flaked stone choppers were found amid a bed of sloth, horse, camel and puma bones. The bones were radio-carbon dated to about 14,200 years and that time frame has been generally accepted.

About the same time people were killing horses and pumas in Peru, Meadowcroft Rock Shelter far to the northeast was also home to ancient, apparently pre-Clovis people. Lying nearly fifty kilometres southwest of today's Pittsburgh in a sheltered, well-watered location, Meadowcroft may have first sheltered humans as early as 16,000 or even more than 19,000 years ago, when the vast Laurentide ice sheet was just to the north. Excavating in the late 1980s, archaeologist James Adovasio found eleven distinct levels of occupation; the earliest, which has been dated at 19,600 years, has met with consider-able skepticism. Later dates—about 14,500 years ago, when the ice was beginning to recede from the area—are easier to accept. The site was used as recently as 700 years ago.

In Canada, what may be very ancient human artifacts have been found in several places, including Calgary, Saskatoon and the Bluefish Caves in Yukon Territory, but none of these has been accepted without question. With Czech archaeologist Jiri Chlachula, Alan Bryan and Ruth Gruhn (of Taima-Taima, Venezuela fame) began digging in 1990 in glacial sediments along the banks of the Bow River. They found more than two dozen flaked stones, similar to others which have been unearthed as far south as Brazil. Despite this, the site has not been generally accepted because the flaked stones were found in a layer of gravelly silt most experts feel is too old, and in the wrong geological context, to bear evidence of human life. And the excavating archaeologists have not yet come up with organic material—bones or plant matter—that can be carbon dated.

Another potentially ancient site was discovered in 1968 in a commercial sand pit in Saskatoon. There, in a layer of sand sandwiched between two layers of gla-cial till seven to ten metres below the surface, a work-man found ancient bones. He alerted the University of Saskatchewan which, despite having only ten days to excavate the pit, found a treasure trove of fossils, including bones of mammoth, horse, camel, bison and deer, as well as small pieces of chipped chert and bone which appeared to have been flaked by humans.

The archaeologists also found a large shoulder blade or scapula bearing many deep and apparently artificial grooves on its flat surface. In Northeast Asia, such scapulae were often used as butcher blocks for cutting up meat. But the most potentially significant find was a metre-long section of mammoth tusk which lay apart from the other bones. A long, thin splinter, which had been split from the tusk and scraped down to the ivory, lay across one end. It's hard to imagine how such a sliver could have been manufactured, except by human hands.

How old were the bones found beneath Saskatoon? Radiocarbon datings for one of the mammoth bones gave a date of 20,000 years, while the two layers of glacial till that sandwiched the sand were dated at approximately 34,000 years for the lower level and 18,000 to 20,000 years for the upper level.

Taken as a whole, and given its location, the site seemed to point to the presence of a very early people with a way of life that was technologically similar to Middle and Upper Paleolithic traditions of the Old World. These Old Stone Age people lived in Europe and parts of Asia between 35,000 and 15,000 years ago; the Canadian site may be just as old, for it appears to have been occupied before the Wisconsin glaciation.

Unfortunately, the sand pit was returned to its owners after a ten-day period of examination and attempts to relocate it have failed.

More widely accepted is the evidence from the Bluefish Caves. Located in the northern Yukon, just fifty kilometres south of Old Crow, the caves today are at the base of a limestone outcropping 250 metres above the valley of the Bluefish River. Fifteen thousand years ago, the caves lay within site of a windswept glacial lake.

The caves were first excavated in 1977 and 1978 by Canadian archaeologist Jacques Cinq-Mars, who found a layer of wind-blown glacial silt, three metres thick in places, overlaid by undisturbed rubble and modern topsoil. It was the kind of situation archaeologists dream about —a well-stratified site, untouched and unaltered. Even better, pollen in the layer of silt proved to be from tundra vegetation, exactly what might be expected during an ice age.

Cinq-Mars began to dig and soon found mammoth, horse, bison, elk and caribou bones, some with cuts and scrapes, clear signs of human butchering. Among the bones were an engraving tool shaped like a chisel, a number of flakes, and tiny, sharp-edged microblades, most made of non-local stone which had been imported into the area.

When the bones were radiocarbon dated, the results met expectations: a mammoth bone which had apparently been butchered was 15,000 years old; the thigh bone of a horse was 13,000 years old. The tools found in association with these bones were given a similar age, but the archaeological community has withheld its final judgement, because a detailed account of Cinq-Mars' work has not been published.

What does all this tell us? Provisionally accepted sites, from Canada's far north to the Peruvian foothills and Chile's southern reaches, are at least 13,000 years old. Unconfirmed sites, from places like Brazil, Pennsylvania and New Mexico, may be much older.

Of course, even accepting the dates for Peru and Chile has fundamentally changed the way we view the peopling of the Americas. For humans to have reached Peru more than 14,000 years ago and southern Chile a millennium later, they must have either entered the continent before the passage between the two great ice sheets closed in Yukon and the North West Territories 20,000 to 25,000 years ago, or come by a different route—perhaps along the coast, as Simon Fraser University archaeologist Knut Fladmark has long proposed. Though still controversial, Fladmark's hypothesis is seen as ever more plausible, thanks to a battery of new studies on glaciation along the coastal regions of British Columbia. Though along some areas of the coast glaciers undoubtedly calved right into the Pacific, many coastal refugia—areas that were unglaciated thanks to protective mountain slopes or ocean currents—also existed.

Perhaps early North Americans came by boat. While this idea has been proposed for decades, it is only recently that it has been taken seriously. Among the reasons for the change of heart are the results of the studies done recently on Kennewick Man, the magnificently preserved 9,400-year-old skeleton that was

Dennis Fast

uncovered along the banks of the Lower Columbia River in July 1996.

Following nine years of political and racial wrangling, in 2005 a team led by forensic anthropologist Douglas Owsley of the Smithsonian Institution finally had a chance to study the bones. They were able to discover a remarkable number of things, among them that he was right-handed; the bones of his right arm were markedly larger than the left, and "so robust that they're bent," said Owsley, likely from a lifetime of spear fishing and hunting. The fishing turned out to be important, for further testing showed that his diet was seventy per cent marine protein, specifically salmon.

The reconstructed skull made it clear that Kennewick Man was not North American, nor was he related to northern Mongolian or Siberian peoples. Instead, he was most like the Ainu, a southern Asian people now found only on Japan's northern islands, but 12,000 years ago lived throughout the coastal regions of eastern Asia. Of even more significance to the question of the peopling of the Americas, the Ainu were completely at home on the ocean. Moving slowly along the coast at the end of the last glaciation, with each generation or two exploring new territory, it's certainly possible that the ancestors of

Kennewick Man could have migrated to a new continent, perhaps without even knowing it.

The other, and likely earlier, North American immigrants were almost certainly the people of northwestern Asia, for a body of dental, genetic and linguistic evidence which seems to conclusively link them to most aboriginal Americans. Studies on the teeth of more than 4,000 individuals, ancient and modern, undertaken by Christy Turner of Arizona State University has led to a series of theories on the settling of the Americas.

What Turner found was that the teeth of most native Americans are closely related to the teeth of the early people of northeastern Asia. He called the distinctive features of these teeth—single-rooted upper first premolars, triple-rooted lower first molars and shovel-shaped incisors—Sinodonty or "Chinese teeth". He found this pattern in northern China as early as 20,000 years ago and believes it may have occurred 40,000 years ago, but it is unknown in ancient skeletons from Europe and the traits are not found in most eastern Asians today. Ancient Europeans and modern Asians largely fall into a category called Sundadont, featuring double-rooted upper first premolars and lower first molars. Nor do present-day Na-Dene (or Athapascan) and Inuit populations, both of which are believed to have arrived in the Americas only in the last few thousand years, share the Sinodont features.

Genetic research backs this theory of an early migration from northern Asia to the unglaciated areas of the Americas. Studies on modern native Central and South American populations showed that more than 14,000 share a particular protein variant in their blood, which has led geneticists to believe there was a single primeval migration of hunter-gatherers deep into the Americas; over the intervening thousands

One can only imagine the wonders that early North Americans found as a new day dawned along the fringes of an uncharted coastline.

of years, these people evolved into distinct cultural groups. The fact that there are more than 1,500 separate languages in South America alone speaks to a great length of time which must have passed since people first settled that continent.

Attempting to determine just how long, scientists have turned to mitochondrial DNA, which is found outside the nuclei of cells and is passed only by the mother, and the Y chromosome, which is passed from the father. Together, they serve as a kind of molecular clock, which changes slowly over time. University of Arizona geneticist Michael Hammer says that mitochondrial DNA indicates the first migration could have taken place 30,000 years ago.

Despite that, the idea that people have been in the Americas for 30,000 years has been difficult for many scientists to accept. In part, that's because until recently it was thought that human beings arrived in Siberia, the accepted launching pad for entry into the Americas, only 35,000 or 40,000 years ago. It was further believed that although Neanderthals, who were known to be well adapted to the cold, were in Europe more than 100,000 years before that, other strains of humanity took much longer to develop strategies to deal with northern climes. Now, those ideas have been supplanted by a growing body of evidence which suggests that humans, in one form or another, have lived and hunted in northern regions for much longer than anyone expected.

Recent discoveries in England have established that ancestral humans were living and hunting on the plains of Sussex a half-million years ago. Years of painstaking excavation have led some to believe that Boxgrove Man—1.8 metres or six feet tall, heavily built and skilled at knapping tools from stone—was *Homo erectus,* a forerunner of *Homo sapiens sapiens,* today's modern humans.

At the same time, another people—dubbed *Homo heidelbergensis*—were inhabiting Germany, and much earlier, perhaps 800,000 years ago, a primitive people may have been hunting in south-eastern Spain.

Archaeologists have long known that *H. erectus* had spread out of Africa as far east as Java as early

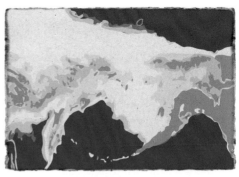

Beringia: exposed between 25,000 and 12,000 years ago

5,000 years ago water levels exposed more of the Aleutian Islands

The Bering Strait region today

These maps show how even small sea level changes can expose land in Bering Strait. In the centre map sea levels are about three metres lower than today.

as 1.8 million years ago. But they felt the ice age, which began about the same time, would have prevented these primitive people from moving farther north. The recent discoveries in Europe indicate this may not be the case. And recently, work in Siberia seems to indicate that humans may have roamed northern Asia for more than 250,000

Early British Columbians were skilled hunters, not only on land but at sea. Bearded seals, right, were among the prey they actively sought.

years. Whether or not this is true, the evidence points to the possibility that human beings may have been poised to cross to the Americas far earlier than we thought.

And how did they cross? Today a narrow strait separates Siberia and Alaska, but during periods of glaciation tiny bands of ice age hunters could simply have walked to the Americas across Beringia, a flat northern plain that was exposed as sea levels dropped by as much as 100 metres. During the early Wisconsin glaciation, between 75,000 and 60,000 years ago, and again during the late Wisconsin glaciation, between 25,000 and 12,000 years ago, Beringia could have provided passage to animals, including humans. During the intervening Mid-Wisconsin interglacial, between 60,000 and 25,000 years ago, Siberia and Alaska were probably separated by a narrow channel of water.

Experts are divided about what the environment was like on this wide, windswept land bridge. Some believe it was thickly vegetated with so-called mammoth steppe or steppe-tundra, stippled with wetland areas where bushes and small trees might have grown, while others feel it was a harsh, sparse environment like today's tundra, but most agree great numbers of enormous animals, collectively known as megafauna, roamed the plain and entered the Americas. And, because hunting such big game was their way of life, humans came after them.

If they came during a period of glaciation—in short, if they walked across Beringia and continued inland east and south—eventually they might have found their way blocked by impassable terrain. Though parts

of Alaska and Yukon were never glaciated and most scientists agree that the great glaciers covering most of Canada may have only coalesced in Alberta for perhaps 3,000 years, for several thousand years before and after that newcomers would have been greeted by an environment that was hostile in the extreme, intimidating even for people accustomed to difficult Arctic circumstances.

But what if the first Americans came earlier, during the mid-Wisconsin interglacial, the period of relative warmth between 60,000 and 25,000 years ago? They might have needed boats to cross the narrow Bering Strait and no evidence of such craft has ever been found. Did they leave their little boats or rafts on what was then the Alaskan shore, an area now submerged beneath the sea? Or perhaps they continued by sea, along the coast, as Knut Fladmark has suggested. The coasts of Alaska and British Columbia were warmer and more productive than other areas, Fladmark believes, even during the height of the late Wisconsin glaciation.

And if they were inhabiting Canada prior to the height of the last glaciation, they would have moved south as the ice did. Are these the people of Pendejo Cave, and Meadowcroft Rock Shelter?

If so, they would undoubtedly have welcomed the great warming and moved north, into southern British Columbia on the heels of their prey. Meanwhile, later emigrants to what really was a New World were moving south along the coast.

Their tracks are everywhere—in the southern Kootenays, along the Fraser River, on the north coast of Vancouver Island—if only we take the time to look.

THE
KOOTENAYS

Displaying one of its many personalities, the Kicking Horse River flows smoothly over its wide valley near Field.

Peter St. John

The Kootenays

Beautiful and, at least to Vancouverites,
charmingly remote, the East and West Kootenays
provide a virtual textbook on many of the things
that set British Columbia apart
from the rest of Canada.
Take plate tectonics, for example.
Like the Rocky Mountains to the east,
the Columbia, Omineca and Cassiar Mountains
(technically, the Omineca belt) were formed
as a result of pressure from collisions
with large island terranes farther west.

Flowing around white sand islands of its own making,
the braided channels of the upper Kootenay River, seen
here in Kootenay National Park's Hector Gorge, tell of
the river's origins high in the Rockies.

The Kootenay region has many charms, including dozens of hot springs. Some, like Halcyon Hot Springs at Nakusp on the Arrow Lakes, right, are developed; others remain just as Mother Nature intended.

For birders, the Rocky Mountain Trench, below, and the Columbia Basin south of Revelstoke draw more than half the species in B.C.

Photos: Peter St. John

AND LIKE THE ROCKIES, the mountains of the Omineca belt are made of rock that has always been a part of the ancient North American continent. The Omineca belt mountains are older than the Rockies, however, and have been through at least two mountain building periods, while the Rockies were built in one major episode about 140 million years ago (see page 14). Nevertheless, locals often refer to the Columbia Mountains (which are comprised of the Purcell, Selkirk, and Cariboo ranges, as well as the western Monashees) as the "Kootenay Rockies".

Much later, between fifty-five and thirty-five million years ago, these "Rocky Mountains of the Far West" were separated from the main ranges by crustal stretching. This rebound effect, which occurred as pressure from the west diminished, created the Rocky Mountain Trench, a deep fissure that is one of Earth's few geo-logical features that can be clearly seen from the moon.

Along the trench south of Prince George, the stretching caused block-faulting, which happens when a huge strip of rock drops downward, creating something like a door opening deep into the earth. Called a half-graben, this kind of faulting can be seen in many places along the trench, but nowhere quite so clearly as in the southeastern Kootenay region, where the Rocky Mountain Trench separates into several valleys and runs south of the American border. The Flathead River Valley, which is actually part of the Rocky Mountains and borders Glacier National Park in Montana, is an example of an almost perfect half-graben valley, with the rock on the west side of the valley precisely matching that on the east. Geologists have determined that the Flathead Valley half-graben is six kilometres deep—that's about two-thirds as deep as Mount Everest is tall. This faulting didn't happen all at once, of course; instead, as the Earth's crust stretched or rebounded as the pressure on the west was relaxed, the rock slipped downward, causing regular earthquakes for millions of years.

The deep fissures that these block-faults created have largely been filled by sediment carried by rivers and streams. The resulting broad, flat-bottomed valley is still impressive, however, particularly when measured against the Rockies to the east. South of Golden, the trench can be easily seen from Highway 95 (see page 55), where viewing spots have been thoughtfully placed at intervals along the highway.

The geology of the Kootenays is also evident in its multitude of hot springs, which occur in clusters; two are in the West Kootenays—one around Ainsworth

Ian Ward

From its deep waters to its soaring slopes, Kootenay Lake—including the West Arm shown below—has provided everything needed for human life for more than 11,000 years.

and the other between Nakusp and Revelstoke, while another cluster is in the East Kootenays, along the Columbia Valley between Cranbrook and Radium. Combined, they create a broad band running south-east to northwest, precisely where the edge of what geologists call the Intermontane Superteranne collided with the margin of the ancient North American con-tinent. The springs follow fractures that pene-trate thousands of metres into the subsurface rock. For more information on how hot springs work, turn to page 60.

The mountain ranges of the Kootenays, and there-fore their valleys and lakes, also have a southeast to northwest inclination, with slopes that face southwest. These, along with the prevailing southwesterly winds, create micro-climates in the valleys that are not only pleasant today—Cranbrook, for example, has more hours of sunshine than anywhere else in the province —but also had a remarkable impact on British Columbia's glacial history and archaeology.

The mountains also divide the East Kootenays, centred around the broad Rocky Mountain Trench, from the West Kootenays. The former has something akin to a continental climate, with hot summers and relatively dry, cold winters. The West Kootenays, by contrast, is a "coastal refugium", ecologically speaking, with mild wet winters and heavy snowfalls that allow typically coastal plants, including western redcedar and salal, to grow. Yet among the coastal refugees, often on the same mountainside, one can often also find out-riders of the dry Interior—including ponderosa pine.

As archaeologist Wayne Choquette has put it, the mountain ranges also create a "solar bowl", a micro-climate that not only melted the great ice sheets much earlier than was the case farther west, but allowed people to live in the region several hundred or even a thousand years before other areas of the province were habitable.

Of course, since melting ice also meant water, most of the earliest archaeological sites have been found on high on the mountainsides on terraces or elevated ancient beaches, for what are still respectably large rivers and lakes today were once much larger and deeper. From west to east, the melting ice created Glacial Lake Columbia (which had first formed as early as 15,000 years ago south of the 49th parallel, along the southern border of the great sheets of ice), Glacial Lake Kutenai, Glacial Lake Windermere, an unnamed glacial lake southeast of Cranbrook and Glacial Lake Elk. Even Glacial Lake Missoula to the southeast, which repeatedly filled as a result of a recurring ice dam and emptied catastrophically at least forty times across the Idaho Panhandle and northern Washington State during the last glaciation, stretched north, at its greatest extent, into B.C.'s Rocky Mountain Trench.

About 12,000 years ago, the glacial lakes began to steadily recede. In the middle Kootenay Valley, charred plant remains almost 11,800 years old have been found, proving not only that an open forest of birch, fir and spruce had been established, but that fire (though whether sparked by nature or humans is not known) was also part of the regional ecology.

Though surprisingly varied ecologically, as the great lakes shrank, the region became linked by three enduring things: the river for which the Kootenays are named; the people who have inhabited its magnificent valleys for more than 11,000 years and the remarkable fish on which those people depended.

Peter St. John

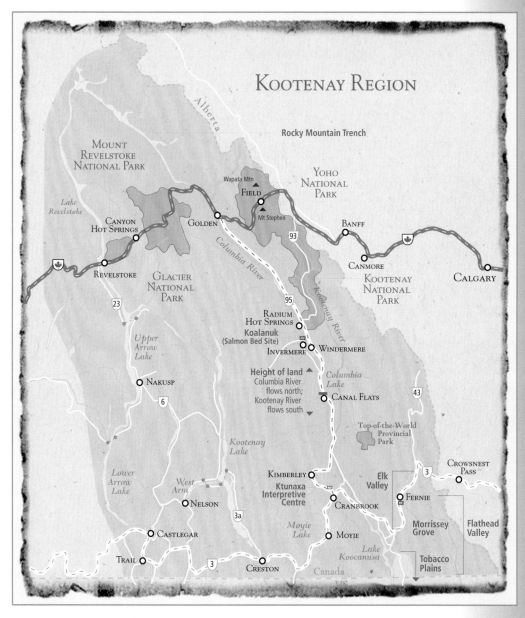

KOOTENAY REGION

Rocky Mountain Trench

Alberta

MOUNT
REVELSTOKE
NATIONAL PARK

Lake Revelstoke

Wapata Mtn
FIELD
Mt Stephen

YOHO
NATIONAL
PARK

CANYON
HOT SPRINGS

GOLDEN

BANFF

93

CANMORE

CALGARY

REVELSTOKE

GLACIER
NATIONAL
PARK

23

Columbia River

95

KOOTENAY
NATIONAL
PARK

Kootenay River

RADIUM
HOT SPRINGS

Koalanuk
(Salmon Bed Site)

INVERMERE WINDERMERE

Upper
Arrow
Lake

NAKUSP

6

Height of land
Columbia River
flows north;
Kootenay River
flows south

Columbia
Lake

CANAL FLATS

43

Top-of-the-World
Provincial
Park

Kootenay
Lake

Lower
Arrow
Lake

West
Arm

NELSON

3a

KIMBERLEY

Ktunaxa
Interpretive
Centre

CRANBROOK

Elk
Valley

FERNIE

3

CROWSNEST
PASS

CASTLEGAR

Moyie
Lake

MOYIE

Morrissey
Grove

Flathead
Valley

TRAIL

3

CRESTON

Lake
Koocanusa

Canada
US

Tobacco
Plains

The Kootenay River is, like the region, and indeed the nations, it unites, a waterway with a multitude of personalities. Beginning as a classic mountain stream, it tumbles south-west through Kootenay National Park on what appears to be a colli-sion course with the Columbia River system. But rather than spilling into Columbia Lake, it curves south, and flows down the Rocky Mountain Trench into what is now Montana, before turning sharply north again.

Re-entering Canada just south of Creston, it meanders over the wide, fertile valley, renewing bountiful marshes that have long sustained both human and animal life, before spilling into a 100-kilometre long fissure in the Earth known as Kootenay Lake. Fed by rivers from south and north, the lake finds its outlet on the west. Flowing down the broad West Arm, the river at last joins the Columbia at Castlegar.

Blessed with these great rivers and lovely lakes, as well as spectacular mountains, significant wetlands and pockets of grasslands, the Kootenay region has a long and fascinating past as well as a promising future.

The Kicking Horse River sleeps under a blanket of snow near Emerald Lake.

Opposite: The banks of the Kicking Horse River, with Cathedral Mountain in the distance.

Yoho National Park

FOR TRAVELLERS along the Trans-Canada Highway, Yoho National Park offers the best of the Rockies in a neat little package. Often described as a hiker's paradise, Yoho (the name reflects an exclamation of wonder in Cree) is perhaps even more exciting for those interested in geology, paleontology, glacial history or the natural environment.

Here are the dividing line between the eastern main ranges and the western main ranges of the Rockies; the magnificent Burgess Shale, described by the late Stephen Jay Gould as "a new world revealed"; the almost equally enthralling Trilobite Beds of Mount Stephen, and just upstream at the headwaters of the Yoho River, one of Canada's most spectacular hanging valleys and one of its highest waterfall. Other wonders include the aptly-named Emerald Lake, with its vivid blue-green water; the nearby Natural Bridge; the hoodoos along Hoodoo Creek at the western edge of the park, and the thundering water of Wapta Falls, where the Kicking Horse River makes a right-angle bend and flows into its spectacular canyon.

Several of these natural spectacles deserve a page (or two or three) of their own. But the park itself is worth a peek … or should that be a peak? Though relatively small in size (1310 square kilometres, compared to Banff National Park at 6641), Yoho includes some of the higher mountains in the Canadian Rockies, including more than twenty that top 3000 metres.

The town of Field marks the dividing line between the eastern and western main ranges of the Rockies.

The eastern ranges are comprised mainly of durable quartzite—extremely hard sandstone—limestone and dolomite, making them less susceptible to erosion than the shaly western main ranges.

Here, too, is the Kicking Horse River. From its rushing upper tributaries, through its wide and tranquil braided midsection to its thundering lower reaches, this is a waterway with many personalities. Along with its primary tributary, the Yoho River, the Kicking Horse was designated a Canadian Heritage River in 1989.

Largely because of its Burgess Shale and Mt. Stephen Trilobite Beds (see pages 44 and 47), the park itself was designated one of seven national and provincial parks in the Canadian Rockies to be jointly listed as a World Heritage Site in 1984. The others are Jasper National Park and neighboring Banff National Park in Alberta, as well as Kootenay National Park, just south of Yoho, and three B.C. provincial parks:

The gatehouse to Emerald Lake Lodge awaits guests.

Peter St. John

Mount Robson and Hamber, both west of Jasper, and Mount Assiniboine, between Yoho and Kootenay.

Originating at the toe of the Yoho Glacier, a tongue of the great Wapta Icefield that lies along the continental divide, the Yoho River is the chief headwaters of the Kicking Horse. Gathering water from its tributaries as it tumbles downstream, the Yoho joins the Kicking Horse (named for an incident in which

geologist James Hector was kicked as he was sur-
veying the region for the Palliser Expedition in
1858) at the Meeting of the Waters just above the
town of Field. Here it slows, wandering through
a wide glacier-carved valley and dropping much
of its load of gravel. The flow freshens as its passes
southwestward through the Van Horne Range.
Near Yoho's western edge it thunders over Wapta
Falls, which are more than sixty metres across,
then turns northwest to roar through a deep
canyon before meeting the Columbia River at
Golden. During its run from the continental
divide to the Columbia Valley, the Kicking Horse
River passes through three major ecoregions of
the Canadian Rockies—Montane, Subalpine
and Alpine.

Jerry Kautz

Main image: The Kicking Horse River flows
under the Natural Bridge—a bridge of its own
making—near Field.

Inset: Traversing Mount Wapta above Emerald
Lake, members of a guided hike to the Burgess
Shale take a breather over lunch.

Dennis Fast

Takakkaw Falls

SPILLING 384 METRES (or 1,260 feet) into the Yoho River Valley, this is one of Canada's highest waterfalls. And its spectacular cascade makes

the idea of a hanging valley easy for anyone to understand.

During the last glaciation, ice covered almost all of British Columbia. In the Rockies, it was deepest in the valleys where, like giant bulldozers, glaciers cleared rock, and widened and straightened what had once been narrow, winding valleys. Smaller glaciers that formed on the heights of the mountains collided with these huge valley glaciers and had their toes continually stubbed. The result was hanging valleys,

where an upper valley abruptly ends at the edge of a much deeper, wider valley below.

This is precisely the action that created Takakkaw Falls—*takakkaw* is Cree for "it is magnificent", according to Father Lacombe's *Cree-French Dictionary* of 1874. The Daly Glacier, part of the extensive Waputik Icefield above the falls, provides the water. Over time, the rapid summer flow has carved a gorge into the top of the cliff, making it seem that the falls appear out of solid rock.

Because it's dependent on the melting glacier for its water, Takakkaw Falls is less exuberant in the morning and swells to full volume on sunny summer afternoons.

Getting There: From the Trans-Canada Hwy, which runs through Yoho NP, turn north on Yoho Valley Road for 13 km to the parking area and follow the short trail to the viewing area.

Takakkaw Falls today, upper left, and in the early 1860s, when painted by Frederick Whymper.

Ian Ward

F. Whymper / British Columbia Archives / PDP-105

41

The Burgess Shale

Linda Fairfield

LOOKING UP AT THE SNOW-COVERED mountains above Field, it's difficult today to imagine life as it was for the animals of the Burgess Shale. Things were very different more than 500 million years ago. Ancestral North America—Laurentia—reclined on its side far to the south, just north of the equator. And what is now Fossil Ridge between Wapta Mountain and Mount Field was at the edge of an undersea cliff in the warm, shallow water along the shore of the ancient continent. Much like Australia's Great Barrier Reef, this spectacular submarine precipice—which we now know as the Cathedral Escarpment, after Cathedral Mountain just east of Field—teemed with life. Though populated mainly by creatures we would not recognize today, this was a community of animals at the cutting edge of evolution.

The Cambrian period was—for reasons scientists are only now beginning to understand (see page 44)—perhaps the most prolific period of phylogeny in Earth's history. And the animals that lived on the edge of the great reef were about to be preserved for all time.

It had taken much of Earth's four-billion-year existence to set the stage for this almost incredible explosion of evolution. For almost three billion years, the plankton floating in the continental shallows had been mostly cyanobacteria (often called "blue-green algae"), tiny single-celled bacteria and cyanobacteria that lacked a true nucleus. They contained chlorophyll, lived by photosythesis and reproduced by cloning, the simple splitting of a cell to create an identical organism. Then, about 2.4 billion years ago, algae with a nucleus—capable of reproducing sexually—developed, and were followed by the first primitive seaweeds. Still, evolution moved with luxurious lassitude. It was not until 750 million years ago that the first very primitive aquatic worms appeared.

About 630 million years ago in the Ediacaran period, more complex organisms appeared. And as the Cambrian period began about 542 million years ago, life began to evolve with startling speed.

But let's backtrack for a moment. Throughout the eons, as the cyanobacteria floating along the shores had lived and died, tiny crystals of lime within their tiny bodies had accumulated on the seabed. Layer upon layer,

Jerry Kautz

Though paleontologists are now beginning to find the ancestors of the animals of the Burgess Shale, this glimpse of Cambrian life forever changed our view of evolution. Inset: *Marella splendens*

they created magnificent beds of limestone. At the Burgess Shale, by 513 million years ago, some of these layers were splitting away and sliding slowly down a gentle slope into deeper water. The broken-off edge of the layers stood more than 150 metres above the sea floor.

At the base of this spectacular undersea cliff—the Cathedral Escarpment —were fine-grained muddy sediments. These had been deposited by currents, tides and waves that were continuously sorting the silt and sand of the seabed. The finest particles, grains of clay less than four thousandths of a millimetre in diameter, were deposited along the foot of the cliff.

Now the stage was set for a remarkable act of preservation. We may never know precisely what happened to the community of creatures that was living at the edge of the escarpment that day half a billion years ago; it may simply have been the collapse of the overhanging edge of the reef into the depths below. Or it may have been an undersea landslide that swept the animals of the Burgess Shale before it. Whatever the cause, the crucial aspect of the day's drama, at least where today's paleontologists are concerned, was that a great many Cambrian animals went over the edge and were buried in the deep silt at the base of the escarpment. The clay, as fine as talcum powder, not only covered all the thousands of creatures completely, gently encasing them as they lay splayed at every angle, so that the feathery gills and delicate legs of

Walcott's "lace crab" (see the following section), as well as the finest details of most of its neighbors were perfectly embalmed, but also invaded every crack and crevice of their internal organs, preserving them forever.

This internal preservation was the most surprising, and by far the most crucial aspect of the fossilization, for it prevented the bacterial breakdown that accompanies the demise of the vast majority of Earth's creatures.

As Murray Cuppold and Wayne Powell write in *A Geoscience Guide to the Burgess Shale*, "… there is more to [fossil preservation] than a simple lack of oxygen, as many bacteria can decompose an animal without it. In fact, almost every animal keeps a supply within its gut; every carcass comes complete with everything necessary to break it down." But in this case, the microscopic grains of clay were forced into even the internal organs of many of the animals, making decomposition impossible and saving the creatures for posterity.

So magnificently mummified were they that scientists have been able to separate their bodies on multiple planes, allowing each fossil to be viewed in layers, rather like an MRI. Today, these fossils are prepared with dental drills that have been adapted for the purpose. Viewed through powerful microscopes, they are allowing new insights into this remarkable community of creatures, from which all later life, including our own species, eventually descended.

Getting There: Those wishing to see the Burgess Shale first-hand must make a reservation for one of many guided hikes that are scheduled every year in July and August. Led by qualified guides, each hike takes a full day and requires that participants be in reasonably good shape. In addition to the regular hikes to Walcott Quarry, other specialized events for children and seniors are also scheduled. Participants are charged a fee. Since the summer schedule fills up fast, it's wise to book early. Bookings are generally taken beginning each year in February. For more information, or to make a reservation, access the Burgess Shale Geoscience Foundation website at www.burgess-shale.bc.ca or call 1-800-343-3006 between 10 a.m. and 3 p.m. Mountain Time.

The Lessons of the Burgess Shale

SINCE THEIR ACCIDENTAL DISCOVERY in late August 1909, the magnificent fossils of the Burgess Shale have opened one window after another on the evolution of complex life on Earth. It was Charles Walcott, head of the Smithsonian Institution, who literally stumbled across the spindly legs and feathery gills of a tiny creature he quickly dubbed the "lace crab", which first hinted at the wonders the Burgess Shale contained.

Walcott, his family and crew had just finished their third field season in the Rockies, studying trilobite fossils and mapping Cambrian formations from Banff to Mount Robson. Travelling by horseback along the ridge trail above Emerald Lake between Mount Wapta and Mount Field, they were halted by a slab of shale that had toppled across the path. As the crew moved to dislodge it, Walcott spotted the faint outlines of a creature he'd never seen before. Delicate and almost impossibly well preserved, it was a glimpse of an ancient world that would profoundly change our understanding of our own. In time, he would name the delicate little animal *Marrella splendens*, in honor of his friend John Marr of Cambridge University, but at the time, he was fully focused on discovery.

Scrambling along the slope that fateful day, he quickly found other animals that were new and strange, as well as familiar trilobites, which he used to date the fossils to the middle Cambrian—by his reckoning about 535 million years ago. (More recent research has updated the duration of the Cambrian period to the interval between 542 and 488 million years ago (mya) and pushed date of the Burgess Shale forward about twenty million years, to about 513 mya.)

The accidental discovery not only changed Walcott's life—he would return to work at the quarry that was soon named for him every summer until just two years before his death in 1927—but thanks to the work of others, the weird and wonderful animals of the Burgess Shale continue to contribute to our growing understanding of the evolution of life on Earth.

Much of that understanding has come in the past thirty-five years, for

Jerry Kautz

Their goal, Walcott Quarry (in the centre of this photo) is in sight for these hikers.

Above and lower right: *Anomalocaris* had paleontologists stumped, while *Opabinia's* five eyes and frontal nozzle astonished Walcott.

after Walcott's life-long love affair with the Burgess Shale, the 65,000 specimens that had been shipped back to the Smithsonian in Washington, along with Walcott's volumes of descriptions, notes and photographs, sat all but untouched for almost a half-century. Perhaps believing that science had learned all the lessons the Burgess Shale had to teach, it wasn't until 1966 that anyone bothered to have another look at the quarry and its collection of remarkable creatures.

That year, Harry Whittington, a professor of geology at Cambridge in England, returned to Walcott's Quarry and with the help of the Geological Survey of Canada, exhumed another 10,000 specimens, which were shipped to Britain. There, with two of his students (and later colleagues) Derek Briggs and Simon Conway Morris, he began cataloguing them. Their interest renewed, the trio also dusted off Walcott's papers and, armed with modern equipment and fifty years of additional scientific knowledge, soon decided that many of the creatures did not fit nearly as neatly into the established biological categories, or phyla, as Walcott had believed they did.

Where to put *Opabinia*, for example, with its five stalked eyes and its long claw-tipped frontal appendage? Or *Hallucigenia*, which Conway Morris believed might have walked on pointed, stilt-like legs? Or *Pikaia*, which Walcott had classified as a worm, but which Whittington could now see had a flexible rod—a nascent backbone?—running down its back? Or *Anomalocaris*, which was found in so many bits and pieces that it defied classification for almost a century?

Slowly, thanks to the fossils' almost unbelievable degree of preservation, and the perseverance of scientists from a range of disciplines, the ancient creatures gave up their secrets. And scientists began to realize that the explosion of life that was the Cambrian had resulted in, to quote the late Harvard paleontologist Stephen Jay Gould, "a range of disparity in anatonomical design never again equalled and not matched by all the creatures in all the world's oceans."

Opabinia was seen to be blessed with both swim fins and legs; was this an early forerunner of the amphibian life that would one day crawl out of the seas to begin to populate the Earth? Or simply a very early multitasker? And thanks to other, better preserved specimens recently unearthed in China, we know that *Hallucigenia* walked on clawed legs, that the pointed spines splayed along its back were armor. *Pikaia*, it seems, was not a worm at all, but perhaps one of our most distant ancestors, with a chordlike spine that might one day evolve into a backbone. The recent discovery in China of a very primitive fish seems to indicate that both it and *Pikaia* had a common ancestor, part of a long evolutionary road life that has led, with all its twists and turns, abrupt endings and renewed beginnings, to all vertebrates and ultimately to human beings.

And *Anomalocaris*, reconstructed at last from the fossilized bits and pieces—one that resembled a headless shrimp, but turned out to be one of two feeding arms, and another that appeared to be a circular jellyfish, but later was recognized as the animal's large, round mouth—was discovered to be the largest predator in the Cambrian seas, a killing machine more than a metre long.

Having finally put all the pieces together and

Walcott's quarry has been deeply excavated in the past century, but there is undoubtedly still much to learn.

Photos: Jerry Kautz

decided on the "what" and the "how" of the Burgess Shale, today's scientists are now focused on the "why". Why, after more than three billion years of evolution that proceeded with exquisite lethargy, did life suddenly explode in a frenzy of biodiversity, unveiling all the body plans we know today—as well as a handful of oddities that seem to have been evolutionary dead ends?

A new generation of specialists—paleobiologists, sedimentologists and paleobotonists among them—have taken up that challenge, aided by newly discovered sites in Namibia, Greenland and China, lightning fast global communication, a body of knowledge that is expanding exponentially and a new interdisciplinary mindset. In Africa are faint signs of the creatures that inhabited the Edicaran period, a previously frustrating twenty-million-year gap in the record of life that immediately preceded the Cambrian. In China are fossils of primeval animals that may well be ancestors of those in the Burgess Shale. And across the board is a willingness to consider the impact of other disciplines on the fundamentals that shaped evolution.

The early Cambrian (which, according to zircon dating, began 542 million years ago) was a time of geological turmoil in what would be Western Canada. Beginning about 780 million years ago, the configuration of the continents changed radically as Rodinia, an early supercontinent, began coming apart at the seams. As the tectonic plates rifted, great blocks of crust tilted, exposing their edges above sea level to form chains of mountainous islands. Mud and rock

eroded from these island ridges was washed down their barren slopes to accumulate as deltas in the ocean shallows. From time to time these deltas would collapse and slide down into deeper water. Volcanic eruptions filled the air with ash. And meanwhile, an ice age was underway.

Yet life survived and prospered. As so often has happened on our remarkable planet, seemingly catastrophic conditions and a rapidly changing environment bumped the pace of evolution.

And just as mammals evolved rapidly to fill the void left by the demise of the dinosaurs at the end of the Cretaceous period, so did a vast array of creatures and body types never before seen appear in the Cambrian. The result of all this was, quite literally, a new world.

Mount Stephen Trilobite Beds

Dennis Fast

Trilobite photos: Jerry Kautz

CLEARLY, given the number and variety found, Mount Stephen was once paradise for trilobites. But then, things were rather different 500 million years ago. What are now the slopes of a mountain in the Main Ranges of the Rockies were then shallow sea beds on the edge of ancestral North America. And this ancient continental core lay along the equator, not in the temperate zone, as much of Canada does today. Even the days during the Cambrian differed; each had just twenty-one hours and a year was comprised of 420 of them, for the Earth revolved faster around the sun than it does now.

Still, to find an entire mountain-side of any creature makes it clear that its environment must have been virtually perfect for survival. And, unlike the remarkable diversity of species found in the Burgess Shale just across the Kicking Horse Valley, the fossils of Mount Stephen are overwhelmingly Cambrian trilobites. Moreover, these "stone bugs", as the off-duty carpenters employed at Mount Stephen House in Field dubbed them when they came upon them on a Sunday in 1886, come in a remarkable range of sizes and shapes. It's as though they'd been called to a trilobite convention.

There are tiny *Agnostida*, swimmers less than a centimentre long that dined on plankton. And medium-sized *Ptychopariida*, which once scurried across the sea floor, snatching tidbits in a low-oxygen environment that apparently didn't appeal to others. There are the predators, including *Ogygopsis* and *Olenoides*, by far the most common fossils in the beds.

In size and shape, *Ogygopsis* have been compared to children's shoe prints. And like those prints, they came in several sizes. Growing by wriggling out of their shells, or exoskeletons, they were up to twelve centimetres in length. In many places on the slopes of Mount Stephen, they can be found gathered head-to-head, as though sharing the latest news, or scattered about a small area, seemingly convened for some important purpose. Surprisingly, *Ogygopsis* is rare almost everywhere else.

The hikers in 1886, or the paleontologists after them, didn't have to look hard to find fossils. As the Geological Survey of Canada puts it, "Virtually every slab on this mountainside boasts complete trilobites." And as authors Murray Coppold and Wayne Powell write, "Trilobites come close to being everyone's favourite fossil."

Easily recognized, perfectly sized, trilobites—so called because their bodies are comprised of three

Above: Alpenglow on Mount Stephen: a fitting tribute to our favorite fossils.

lengthwise lobes—are also popular with paleontologists. The reason? The thousands of forms that developed between the early Cambrian 542 million years ago and the end of the Permian 291 million years later, allow the rocks in which they appear to be securely dated.

The fossils on Mount Stephen are from the early trilobite tenure and for the past century have been increasingly popular with visitors to Field and Yoho National Park. Alas, that is no longer the case. In 2005, responding to damage to the fossil beds caused by increasing traffic, the beds were closed to the public.

The Rocky Mountain Trench

THIS IS ONE OF THE EARTH'S few geological features that can be seen from the moon. Located just west of the Rockies, the Rocky Mountain Trench is actually a system of valleys created by two quite different phenomena. The southern section, which runs from the B.C.-Montana through eastern British Columbia to Prince George, was created when the Earth's crust stretched across B.C.and broke along a huge fault line about forty-five million years ago.

The northern section, from north of Prince George to the Liard River, is also marked by a major fault, but instead of moving downward (as the southern section did), here the western side has moved northward, like the land on the west side of California's famous San Andreas Fault. Between three and sixteen kilometres wide, the trench varies in altitude between 600 and 900 metres above sea level.

Bounded on the east by the main ranges of the Rockiy Mountains, and on the west by the Columbia Mountains, the flat-bottomed trench lies between steep mountain walls.

Though geologists are still debating the various forces that contributed to its origins, it seems that the trench is a result of nearly 140 million years of geological push-pull and lateral movement along the ancient margins of North America. British Columbia was formed when a series of volcanic island terranes— rather like today's Phillipines—that were riding the edge of the undersea Pacific plate slammed obliquely onto the west coast of North America. Beginning about 170 million years ago, in the middle of the Jurassic period, these volcanic islands began bumping and scraping along the continental margin in a series of lengthy and violent collisions.

The continent acted like a wedge, peeling the lighter rock of the volcanic islands off the underlying oceanic plate and adding them to its already enormous territory. Meanwhile, inland from the collision zone, the new territory served as an enormous bulldozer. Rumpling the sedimentary layers northeastward, it created first the Columbia Mountains and then the western part of the Rockies. At the time, the two mountain ranges were one: the Rocky Mountain Trench had not yet formed to divide them.

Out in front of the growing mountain range was a trough in the Earth's crust. Fringing it were thick swamps, muddy deltas and shallow bays, home to many species of dinosaurs. Time and pressure transformed these ancient swamps into some of the world's most bountiful coal fields; today they stretch from southeast of Fernie to north of Sparwood and into the Crowsnest Pass.

North America's relentless westward movement continued; about eighty-five million years ago, the continent began to close with another series of volcanic islands. As the Coast Mountains, Vancouver Island and Haida Gwaii were cemented onto the continental edge, the pressure sent the Columbias and the main ranges of the Rockies soaring skyward to heights comparable to today's Himalayas. The trough lying east of the ranges buckled upward, raising the soft layers of sandstone, shale and coal to form a long strip of hilly highland lying at the elevation of Tibet. Over time, erosion stripped these soft sediments away, exposing harder layers that weathered into the front ranges of the Rockies.

Now, finally, British Columbia's outlines were complete and, about fifty-five million years ago, the pressure on the continental margin began to diminish. Just as a Slinky toy expands when the pressure is released,

Peter St. John

Dennis Fast

The trench (opposite) can be best seen at Golden, and at several viewpoints along Highway 93/95.

the land between the Columbia Mountains and the Rockies stretched until in several places the crust split, faulted blocks of it slipped downward. When this happens without any sideways motion, and the rock on one side of the valley matches the layers of rock higher or lower on the other side, geologists call the process "half-graben block faulting", where the Earth's crust behaves like a door opening downward. This half-graben was the Rocky Mountain Trench, which separated the Rockies on the east from the Columbia Mountains on the west.

The MacDonald Range, which is just north of the U.S. border, marks where the Rocky Mountain Trench splits into two valleys. The more easterly of these, the Flathead Valley, is particularly notable. Not only is this a half-graben six kilometres deep, but this thinly populated region supports the largest grizzly bear population in the North American interior and is one of the few places where grizzlies are still found in a grasslands environment. It also boasts a large number of eagles. For these and other reasons, it has been proposed as a new national park. Bordered on the south by Montana's Glacier National Park and on the east by Alberta's Waterton National Park, the proposed park (tentatively named the Flathead National Park Reserve) would significantly increase the size of the land under protection and increase tourism opportunities in southeastern B.C.

Though found mainly in the mountains today, grizzlies were originally grasslands creatures. This mother and cub are therefore right at home.

With B.C.'s mountain building largely complete, a long period of erosion began, as wind and water wore down the landscape. Ancestral rivers carved winding courses along the trench and a series of glaciations (see page 16) over the past two million years, straightened the walls and widened the floor, creating the remarkable feature that can be seen from space.

Nine of B.C.'s rivers—including the Columbia, Fraser, Thompson and the Kootenay—have their upper reaches in the trench. Most exit the valley through canyons.

At the end of the last glaciation, much of the southern Rocky Mountain Trench was intermittently filled with water from a northerly tongue of Glacial Lake Missoula in northwestern Montana. Repeatedly blocked by ice dams to the west, the impounded water of Lake Missoula rose and then dramatically drained across eastern Washington and western Oregon an estimated forty times between 15,000 and 13,000 years ago. At its highest, Lake Missoula threatened to spill over the Crowsnest Pass, more than 1350 metres above sea level.

To the west, the Purcell Trench (which holds today's Kootenay Lake) and the Selkirk Trench (where the Columbia River widens into the Arrow Lakes) also held large, deep glacial lakes. Terraces of lake silt and wave-cut shorelines were left high above the valleys in many places, such as the confluence of the Elk and Kootenay Rivers (see page 83), and the meeting of the Kootenay and Columbia Rivers. These ancient benches were perfect campsites for the early inhabitants of the valleys, and have proven to be bountiful sites for archaeologists (see The Ktunaxa on page 67).

49

Mount Revelstoke & Glacier National Parks

TRAVELLERS PASSING THROUGH Glacier National Park and along the southern edge of Mount Revelstoke National Park just to the west may not realize that this is an environment found nowhere else on Earth. Here, in the Selkirk Mountains, as well as the Monashee and Cariboo Mountains to the north and the Purcell Mountains to the south, are the last old-growth remnants of the world's only inland temperate rainforest. Biologists calls these "antique forests", for they have been regenerating themselves since the last glaciation and exist thanks to a unique combination of geology and climate.

Glacier National Park lies in the Columbia Mountains, the wider range that includes the Purcells, Selkirks, Cariboos and Monashees.

The Columbias are geologically distinct from the Rockies, but—as described on page 14—closely related.

Peter St. John

Awaiting the spring freshet, the Columbia River weaves through and around mid-stream sandbars at Revelstoke.

The rugged mountains seen along the Trans-Canada Highway and the railway are comprised mostly of quartzite, slate and other forms of metamorphic rock dating back more than 500 million years. All were thrust skyward in the tectonic plate collisions tht built Canada's western mountains beginning 140 million years ago. Subsequent erosion by rivers and glaciers has carved the sharp, angular peaks and steep-sided canyons of the Columbias, displayed so impressively around Rogers Pass.

GLACIER NATIONAL PARK

Beaver River

Hermit Mt

Mt Rogers

Rogers Pass

MOUNT REVELSTOKE NATIONAL PARK

Mt MacDonald

Columbia River

Mount Sir Donald

Mt Bonney

Mt Jupiter

Inverness Peaks

Albert Canyon Hotsprings

Mt Dawson

Mt Revelstoke

Illecillewaet River

Revelstoke

Steep-sided, with sharp, angular peaks, large cirques and deep snow, the Southern Selkirks are increasingly popular among backcountry skiers and, come summer, with hikers and climbers.

Peter St. John

Since B.C.'s weather almost invariably originates over the Pacific Ocean, year-round these mild wet, westerly air masses flow east, unimpeded, until they hit the wall of the Columbias. The result is heavy summer rain and deep winter snow, particularly at higher elevations, where up to twenty-three metres may fall over a winter. Little wonder that the two parks, though not particularly large, have 400 glaciers within their boundaries.

The natural effect of this copious year-round precipitation is a primeval forest crowned with huge western hemlocks and giant western redcedars, including some that are almost 2,000 years old. Here, too, are more tree species than are found together anywhere else in B.C., for this remarkable ecosystem brings together plants from the coastal, southern interior and boreal forests.

Not surprisingly, perhaps, these old-growth forests have long been home to a huge diversity of wildlife— among them mountain caribou, now on Canada's endangered list (see page 91); mountain goats and bighorn sheep; grizzlies and wolverines—both considered vulnerable; black bears and cougars, as well as many species of birds that depend on cavities in mature trees for nesting sites and raising their young.

An old-growth forest is also better able to survive natural disasters, including fires triggered by lightning, and invasions of insects, like the mountain pine beetle that is currently sweeping through forests of lodgepole pine in the B.C. Interior and has caused the closure of one of the main campgrounds in Glacier National Park. In fact, studies of old-growth forests demonstrate that invasions of bark beetles are a regular occurrence and one that a forest of different species and a variety of ages can withstand.

Despite all this, and though British Columbia's old-growth inland temperate rainforest is unmatched anywhere on Earth, most of it has disappeared in the past half-century. More than two million hectares have been logged since 1960 and at the lower elevations, only 3.2 per cent of old-growth forest remains.

Within the boundaries of Mount Revelstoke and Glacier National Parks, the forests can no longer be logged, though much of the forest in both parks is maturing second-growth forest, largely because fires burned large tracts of forest during the construction of the CPR railway in the 1880s. Both parks do have prime examples of old-growth forest, however, and have created easily accessible trails, the Giant Cedars Trail near the eastern boundary of Mount Revelstoke NP and the Hemlock Grove Trail near the western edge of Glacier National Park. But neither park is

51

Peter St. John

Peter St. John

The Giant Cedars Nature Trail, above, meanders through an old-growth rainforest unmatched anywhere on Earth. Above the treeline, alpine meadows blaze with paintbrush, at right.

treeline, creating open meadows that explode into a brief blaze of color in August, with paintbrush, lupines and monkey-flowers. Fully half the park is above the treeline and much of this is rock, permanent ice and snow or tundra. Grizzlies, which also are found at the lower elevations during berry season, mountain goats, white-tailed ptarmigan and caribou are among the many animals that summer above the tree line.

large enough to provide any real protection for the forest ecosystem as a whole, or for many of the animals at risk, particularly those, like mountain caribou and grizzlies, that must range widely for food.

Over the past thirty years, the plight of disappearing old-growth rainforests on B.C.'s Pacific coast has been spotlighted with positive results on the part of the provincial government. But the province's even rarer inland rainforest has been virtually ignored.

Each of the two national parks encompasses examples of the best of the Columbia Mountains. Mount Revelstoke NP, which is entirely within the Selkirk Range, stretches north and east from the outskirts of Revelstoke and provides a snapshot of the "Rainforest, Snowforest, No Forest" ecosystem of the region. Driving up into the park on the Meadows-in-the-Sky Parkway, the rainforest, described above, gradually gives way to the "snow-forest", dominated by Englemann spruce and fir. This forest thins near the

Ian Ward

Glacier National Park is more extensive and higher than Mount Revelstoke. And not surprisingly, glaciers are major features of the park, which is dominated by ten peaks ranging from 2600 to 3390 metres. By comparison, the highest peak in Mount Revelstoke National Park is Mount Coursier at 2646 metres.

Glacier NP is famous for Rogers Pass, which is crossed by both the Trans-Canada Highway and the Canadian Pacific Railway line, and bisects the mountains of the Northern and Southern Selkirks. The latter include Illecillewaet Glacier, described as a "must-see" destination for more than a century. The Great Glacier Trail is one of several trails that begin at the end of the road to Illecillewaet campground. It involves a moderate climb and a ten-kilometre return hike. Branching off the trail are others to other peaks (such as Avalanche Crest, one of the most dramatic viewpoints in the region) and glacers in what's called the Mount Sir Donald Group.

Glacier National Park, which was created in 1886, is considered the "birthplace of mountaineering" in Canada and is still a popular place for climbers. It is also famous for its outstanding back-country skiing.

If you plan to go, purchase one of the excellent books on climbing in the region (some of the best are listed in the bibliography at the end of this book), as well as good maps from the Geological Survey of Canada outlets in Calgary or Vancouver, specialty stores or the Visitor Information Centre at Rogers Pass. And never climb alone.

Getting There: The elevations in Glacier NP mean a short hiking season, so hikers should contact the Visitor Centre in the shoulder seasons (April, May, September and October) for information on conditions. The park's two rustic campgrounds are open from June to October. At Mount Revelstoke, the Meadows-in-the-Sky Parkway, and the Eva Lake and Jade Lakes trails above it, open over a series of weeks beginning as soon as snow is gone from the lower elevations. As the snow melts, the road is opened (during the day only—check for seasonal hours) to higher elevations, with the summit finally snow free by the third week of July.

Mount Sir Donald, at 3284 metres among the highest in the Selkirks, guards Rogers Pass.

Ian Ward

The iron-rich rock at Radium Hot Springs
glows copper in the morning light.

The Columbia Valley

The Columbia Valley holds the headwaters
of the Columbia River, a stretch of the upper Kootenay,
some of B.C.'s most productive wetlands
and a half-dozen hot springs.

FOLLOWING THE broad, flat valley of the southern Rocky Mountain Trench (see page 48), the valley of the upper Columbia River has become almost synonymous with the trench itself. Yet the two should be considered separately.

If the trench, the deep cleft in the Earth that has been largely filled by sediments from the surrounding mountains, can be seen as the cradle, the Columbia River, with its headwater lakes, sinuous upper reaches, wetlands and marshes, is the region's main fount of life. Here is one of British Columbia's most vibrant ecosystems, home to hundreds of species of birds and most of the province's large mammals.

Here, too, are hot springs in abundance. Though sediments have buried the basement rocks in the valley bottom, reminders of its intimate acquaintance with the Earth's molten core percolate up through the rock at its edges. Some are the focus of lovely resorts, while others remain beautifully *au naturel*; as a result, there is something here for almost everyone.

Radium Hot Springs, almost halfway between Golden and the Crowsnest Highway, anchors Kootenay National Park. Like Yoho to the north (see page 39), the park is one of a quartet of national parks that were jointly named a World Heritage Site in 1984.

Kootenay NP includes Vermilion Pass, which has long provided access through the Rockies. And for millennia, the grasslands of the upper Kootenay Valley were home to herds of elk, deer and bison and, on their heels, early hunters. Today, the Kootenay Parkway (Highway 93) links the Trans-Canada Highway north of Banff with Highway 95, which follows the Columbia Valley, providing an alternate route through the Rockies.

For thousands of years, the Columbia Valley was the preserve of the Ktunaxa people (see page 67) and also, in more recent centuries, Kinbasket Shuswap people from the northwest. Each autumn, both eagerly awaited the chinook salmon who made the long and tortuous 1900-kilometre journey up the Columbia

Peter St. John

Fire, whether ignited by man or nature, is part of maintaining a healthy forest. This blaze, captured in pen and ink by John Pedder, occurred in 1887.

John Pedder, National Archives of Canada, C-103963

River from the Pacific. Ktunaxa camps have been found up and down the river, including Koalanuk—"where the lake empties into the river" (see page 65). Here, at the north end of Windermere Lake, just a stone's throw from the modern community of Invermere, more than 1,000 years ago the Ktunaxa came here to greet the spawning salmon. But they did more than process fish here; archaeologists excavating at what they call "the Salmon Beds Site" also found the bones of bison, white-tailed deer, elk, black bear and beaver. Clearly, this was a harvest camp, a bountiful place where many types of food were processed for the coming winter.

With a history in the region that goes back more than 11,500 years, the Ktunaxa are perfectly at home throughout this region. There is hardly a mountain peak or a lakeshore that does not bear signs of their distinctive presence. From Top-of-the-World Provincial Park (see page 76) southeast of Skookumchuck, where quarries in the chert outcroppings have been found in the region's ancient Orodvician limestone, to the sparkling St. Mary River—one of B.C.'s premier trout streams—where fishing camps are scattered along the shores, hundreds of sites throughout the Columbia Valley and the rest of the Kootenays speak of their long history here. Today, that story is eloquently told at the Ktunaxa Intpretive Centre on the St. Mary River (see page 71).

So, though many equate the Rocky Mountain Trench with the Columbia Valley, the two are not the same. The trench continues north to the Yukon border, far north of the river's most northerly reaches, and stretches south of the Columbia River headwaters as well. The southern reaches of the trench are occupied by the Kootenay River (see page 72). Flowing out of Kootenay National Park on a course that seems destined to intersect with the Columbia, the Kootenay instead turns abruptly south down the trench, missing the headwaters of the Columbia by a whisker.

The final pages of this section focus on the white sturgeon (see page 74), one of two piscine species that, long ago, made a remarkable transition from a long history as anadromous or ocean-going creatures to spending their entire lives in fresh water. The story of the other, kokanee salmon, can be found on page 121.

The Columbia River

Peter St. John

COLD MOUNTAIN-FED LAKES, a sinuous wetlands stream, serpentine waterways, thundering falls, deep gorges and a vast mouth to the sea: all these are the Columbia River. Almost 2000 kilometres long, shared between Canada and the United States, the Columbia drains an area larger than France, ranking sixth in North America in terms of its watershed.

Along its route, the Columbia nourishes one of the most internationally significant wetlands on the globe, creates riverine highways through some of the most inaccessible regions of southern B.C. and, because its lower reaches were south of the edge of the ice during the last glaciation, provided an early route for the aquatic repopulation of British Columbia's waterways.

Meandering along a deeply silted basin between the Rockies to the east and the Columbia Mountains on the west (today sometimes called the Valley of a Thousand Peaks), the

Dennis Fast

upper Columbia is unrecognizable as the mighty river that eventually empties into the Pacific on Oregon's northern border. It begins in Columbia Lake, just north of Canal Flats and is separated from the upper Kootenay River by less than two kilometres. Fed by cold mountain streams—including the Kicking Horse and Blaeberry Rivers—it winds down the nearly flat block faulted basin of the southern Rocky Mountain Trench (see page 48), widening into Windermere Lake and dozens of shallow ponds, from Canal Flats northwest to Donald—more wetlands than waterway. These are the remains of the long, glacial lake that once filled the valley.

This broad, lush section is globally important, for its interconnected system of bulrush swamps, sedge meadows and shallow lakes is a piece of paradise for many species. Here is Western Canada's second-largest colony of great blue herons, as well as bald eagles, hawks, loons, trumpeter swans and more than 100 species of song birds. In fact, sixty-one per cent of all birds in Canada can be found in the Columbia Basin, which includes the lower Columbia region downstream of Revelstoke, between the Columbia and Monashee Mountains.

Relatively warm and dry, compared to the valleys of the Eastern Rockies, edged with forests of aspen, cottonwood, willow and white spruce, the Columbia

Above: South of Revelstoke, the Columbia widens through Upper Arrow Lake.

Below: The Blaeberry River rushes over its rocky bed to join the upper Columbia near Golden.

Valley wetlands have also long served as important winter habitat and a crucial travel corridor for elk, moose, caribou and grizzly bears. Seventy-four per cent of all terrestrial species in B.C. are found in the Columbia basin, though twenty per cent of those— a total of eighty-four species—are currently red- or blue-listed, considered endangered or threatened.

The upper reaches of the Columbia, along with the upper Kootenay, are also home to several unique forms of cold water fish species, including char, trout and sculpins. All populated the river between 13,000 and 9,000 years ago, when higher water levels and ice dams created a series of glacial lakes with connecting channels. These were later closed to fish by impenetrable waterfall barriers or drainages that were no longer linked (see the Thompson River on page 103).

In fact, the fish of the Columbia Basin are among the most distinct in North America. Scientists believe this is due to the repeated glaciations over the past two million years, which intermittently swept British Columbia clean of its fish populations. Because the lower Columbia was south of the glaciated area, and because several sections of the Kootenays, including the headwaters area in the lower Columbia Valley, were unglaciated much earlier than elsewhere, the forms of char, trout and sculpins that found their way to the river's upper regions were genetically different than their later relatives, which populated the other rivers of southern B.C.

Remarkably, given its enormous distance from the sea, over time the upper Columbia also became an important spawning site for salmon. Until the last century, which saw the construction of huge dams on the river in both Canada and the U.S., the Columbia was home to the world's largest runs of chinook salmon, as well as spawning runs of other salmon species.

For thousands of years, huge chinook travelled more than 1900 kilometres from the Pacific to spawn in the gravelly shallows along the river's upper reaches. And their regular arrival, between late August and early October, drew the Ktunaxa people (see page 67), who have lived in the Kootenays for more than 11,000 years.

Here too, testament to the continued underlying geological activity (see page 60), are hot springs that have refreshed and healed people for millennia.

In May of 1846. artist Henry James Warre painted this view looking east toward the Rockies from the Boat Encampment on the Columbia. Here, canoes awaited the west-bound fur brigades arriving from Jasper.

The ferry across Upper Arrow Lake.

Continuing north along the Rocky Mountain Trench, growing as it is fed by streams from the Columbia and Clemenceau Icefields, the river turns sharply at Mica Creek, to flow south again past Revelstoke. Filling a deep cleft in the earth between the Selkirk and Monashee ranges, it forms Upper and Lower Arrow Lakes, before being joined by the Kootenay River at Castlegar, just north of the U.S. border.

In Washington State, enlarged by the Okanogan River (the spelling is different south of the border), the Columbia crosses the Channeled Scablands and finds the gorge created by the massive Missoula or Bretz Floods. These massive outpourings were created by the repeated failures of the ice dam that created Glacial Lake Missoula at the end of the last glaciation. Geologists estimate that Lake Missoula formed and reformed about forty times over a period of about 2,000 years between 15,000 and 13,000 years ago. Each of the cataclysmic floods created walls of water as high as 400 feet and hurled boulders and icebergs seaward at speeds of more than 100 kilometres an hour.

Finally, mixing with the tidal waters of the Pacific, the mighty Columbia reaches the sea at Astoria, Oregon, just west of Portland.

Getting There: To reach the headwaters of the Columbia River at Columbia Lake in the Rocky Mountain Trench, take Hwy 95 south from the Trans-Canada Hwy at Golden. To follow the river downstream, head north of Golden on the Trans-Canada to Donald, where the highway crosses the river and Marl Creek Park offers an opportunity to camp next to it. Kinbasket Lake, a widening of the river that stretches far to the north, can be reached by turning onto Columbia West Forest Service Road just before the Trans-Canada turns south to enter Glacier National Park (see page 50). Crossing the Columbia Mountains and following the Illecillewaet River, the Trans-Canada is reunited with the Columbia at Revelstoke. To continue following the river, take Hwy 23 to Nakusp, crossing the north end of the Upper Arrow Lake by ferry from Shelter Bay to Galena Bay. The main highway (No. 6) then leaves the river, heading west and south along the Slocan River and Slocan Lake to join the Redcoat Trail (Hwy 3) at Castlegar. Hwys 22 and 22A follow the river to the Washington border.

Dennis Fast

Hot Spots & Hot Springs

ONE OF THE CLEAREST indications of British Columbia's birth by fire is its wealth of hot springs. According to the second edition of *Hot Springs of Western Canada*, by Glenn Woodsworth, of the approximately 110 hot and warm springs in Canada, the majority—about eighty-five—are in B.C. And locating them on a map of the province makes it clear they are concentrated along lines of tectonic activity.

Since the heat that warms the water in all natural hot springs comes from deep within the Earth, and since that water needs relatively easy access both down to the heat source and back up to the surface, it follows that tectonically active areas—those with fractures and faults—are where hot springs are found. The islands of Japan, for example, which perch along the western edge of the ever-volatile Pacific Plate, are riddled with 4,500 hot springs, according to Woodsworth.

In Canada, the East and West Kootenays are particularly blessed with an abundance of steaming springs. Here, dotted along the valleys or perched on the slopes, is something for literally everyone: the up-scale resorts at Radium and Fairmont Hot Springs in the Columbia Valley and at Nakusp and Halcyon Hot Springs on Upper Arrow Lake; the hot and steamy "caves" at Ainsworth Hot Springs on the west side of Kootenay Lake; the wonderfully accessible pools at Canyon Hot Springs on the Trans-Canada Highway east of Revelstoke,

Photos: Peter St. John

and the region's many secluded natural springs such as those at Dewar Creek in the Purcells northwest of Kimberley.

It makes sense that the Rocky Mountain Trench has more than its share of springs and that some of them, including Dewar Creek and the upper (and minimally developed) springs on Emanation Hill above Fairmont, are quite hot, for though deeply filled with sediment, the trench is quite literally an open door to the Earth's interior, caused by crustal stretching and block faulting about fifty million years ago.

The faults under this part of the province are up to six kilometres deep, providing conduits up which water can move. The Earth's temperature increases approximately 30°C with each kilometre in depth, mainly as a result of the continuous decay of radioactive elements, particularly uranium and thorium. In most places, the Earth's subsurface temperature reaches the boiling point of water between three and four kilometres down. If water, running downhill, can follow a fault to a depth where it is heated, and then—because hot water, like hot air, is less dense than its cold equivalent—find another fault to rapidly return to the surface,

Top: Relaxing in Lussier Hot Springs near Skookumchuk; upper left, Silver Springs, not far from Fernie, and left, water from this spring, near Albert Canyon, is piped almost two kilometres to Canyon Hot Springs.

Photos: Peter St. John

Inset: The old baths at Fairmont Hot Springs sit on a hill of tufa, while the lower spring near today's pool supports a lush growth of algae as it flows into Fairmont Creek.

a convection system is created that will continuously supply a hot or warm spring.

A spring's outlet temperature largely depends on the length of time the water takes to rise to the surface, though such things as surrounding cold streams can also cool the water as it exits. And molten rock, found in volcanic regions such as Yellowstone or Iceland, has enormous heating capacity, enough to not only produce natural hot springs, but also to provide geothermal energy, as is the case in many tectonically active areas, including Japan and California. The Meager Creek Springs, at the base of

Mount Meager north of Vancouver (see page 208), are being tested for just this purpose.

The composition of spring water is dependent on the type of rock through which it passes; layers of limestone imbue water with calcium, and often create extensive tufa deposits at the outlet of the spring.

The Ktunaxa undoubtedly knew the existence of virtually all the springs in the region, for there is almost nowhere in the Kootenays they didn't camp or hunt or fish over the millennia. Moreover, it's very likely that many of the springs drew wildlife to lick the minerals deposited there. It's even possible that a hot spring in the Kootenays may have drawn a rheumatic grizzly for a therapeutic soak, as the Cave and Basin Hot Spring purportedly did in what is now Banff National Park.

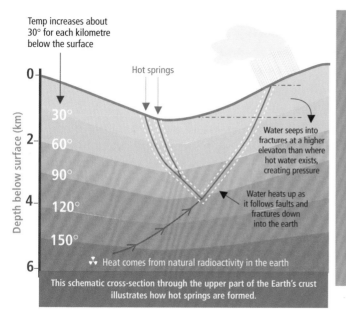

Temp increases about 30° for each kilometre below the surface

Hot springs

Water seeps into fractures at a higher elevaton than where hot water exists, creating pressure

Water heats up as it follows faults and fractures down into the earth

Heat comes from natural radioactivity in the earth

This schematic cross-section through the upper part of the Earth's crust illustrates how hot springs are formed.

Getting There: Every spring is different, not only in size, temperature and composition, but in facilities and accessibility. Information, particularly on springs that have been developed commercially, is available on the internet. There are also several good guides to B.C.'s hot springs available, including the one mentioned above. Keep in mind that you should never bathe or swim alone, and should you feel faint or dizzy, get out immediately and cool down. Keep a particularly close eye on children, the elderly or anyone suffering from high or low blood pressure, or heart or circulatory problems; they may easily overheat.

Kootenay National Park

STRETCHING SOUTH from Deltaform Mountain, which rises more than 3400 metres into the sky on the Alberta border, to Dry Gulch Provincial Park just south of Radium Hot Springs, Kootenay National Park encompasses everything from glaciers to cacti. Its climate ranges from cool and moist in the north to relatively dry in the south, the latter thanks to the rain shadow effect of the Columbia Mountains to the west. This range creates an environment that has long suited many animals, particularly ungulates. Today, these range from mountain sheep, which can be seen on the

found. Similar both to a point unearthed at the Lehman Site, in Shuswap territory near Kamloops (see page 106) and to others found at sites along the Pend D'Oreille River near Nelson, it is believed to be an atlatl point between 3,500 and 6,000 years old. This not only demonstrates the reach of a particular technology, but may also point to the long association between the Ktunaxa and Shuswap people.

Looking southwest up the Simpson River reveals Simpson Ridge, at left, in Mount Assiniboine Provincial Park, and Assiniboine Pass through the Rockies in the centre of the photograph.

Peter St. John

cliffs above Radium Hot Springs; white-tailed and mule deer, which graze along the Kootenay Park-way (Highway 93) with such regularity that park officials are embarking on a variety of methods to avoid collisions, and elk, which populate the grasslands along the Kootenay River Valley. In the past, bison also ranged over the valley grasslands. Little wonder that this bountiful region was repeatedly used as a hunting ground for at least 6,000 years.

Among the many hunting camps excavated by archaeologists is one on a high terrace above the Kootenay River, where a black siltstone projectile point with deep corner notches and a straight base was

But it was not just the valleys that attracted early hunters. At Kaufmann Lake, a seemingly isolated mountain region at the northern end of the park, Parks Canada archaeologist Rod Heitzmann excavated a quartz crystal workshop, where carbon from a hearth yielded a date of 4,470 years BP. Other sites in the park include a grassy col just below the Sinclair-Kindersley Pass near the south end, where bone samples proved to be about 1,800 years old. And along the

Kootenay River, a side-notched projectile point—similar to arrowheads being used on the Alberta plains—proved to be of Top-of-the-World chert (see page 76). Blood on a stone knife found near the arrowhead was bison; it was dated to 380 BP. Taken together, this evidence shows that bison were still found west of the Rockies until quite recently and that the Ktunaxa (for it was likely they who quarried the chert) had adopted bow and arrow technology.

Weapons technology was not the only thing that was changing as the centuries passed. Recent sediment cores by ecologists Robert Walker and Douglas Hallett have shown that changes in climate created significant environmental changes, as well. To obtain a long view of those changes, Walker and Hallett took sediment cores from Dog Lake, a small lake just northeast of the McLeod Meadows Campsite in the Kootenay Valley. One allowed them to analyze the pollen and other small fossils, as well as charcoal particles, at approximately forty-year intervals over the past 10,000 years. A smaller core was used to paint a more precise picture of the environment over the last thousand years.

Today, the forest around the lake lies in the Montane Spruce biogeoclimatic zone, a narrow band of dense woodlands that nestles between the high-elevation subalpine forests of spruce and subalpine fir and the open lower-elevation forests of Douglas-fir or lodgepole pine that dominate B.C.'s dry southern interior. But that was not always the case. The scientists found their 10,000-year record fell rather neatly into three periods: Period 1, the last four thousand years; Period 2, the period between 4,000 and 8,000 BP and Period 3, from 8,000 to 10,000 BP, following the retreat of the glaciers. Of these, Period 2 (which climatologists know as the Hypsithermal) was clearly much warmer and drier than the others. Featuring dry open forests of ponderosa pine, large amounts of charcoal showed that this was a period of frequent forest fires and low lake levels.

The last 4,000 years were generally wetter and cooler, with higher lake levels and fewer fires. Within that period, however, core sediments for the last 1,000 years verified what climatologists tell us to be true. During the Medieval Warm Period, four centuries of consistently warm temperatures in the Northern Hemisphere between about 900 and 1300 AD, droughts occurred regularly and fires repeatedly destroyed the forests surrounding Dog Lake. Moreover, it seems that infestations of bark beetles (like the one currently rampaging through the pine forests of southern B.C.) were more common during periods of drought.

Elsewhere, during the same period of global warming, the Norse settled Greenland and established farms, while the Scots planted vineyards north of Hadrian's Wall. But this lengthy period of global warming was not to last, for hard on its heels came the Little Ice Age, which lasted for almost 500 years. This was the period that saw huge increases in the size and number of glaciers in the Rockies, a

Photos: Peter St. John

The Ochre Beds along Ochre Creek provided iron-rich red earth used by many cultures to paint everything from tipis to clothing and even faces.

to recording visions on rock. It was also traded widely.

At the south end of the park is Radium Hot Springs, one of many natural hot springs in the Kootenays that have long drawn people to their soothing waters.

frigid end to the Greenlandic settlements, repeated crop failures in the Scottish vineyards and even regular public skating on the Thames in England. At Dog Lake, deeper lake levels and wet, closed forests meant far fewer fires.

Kootenay National Park has other things of interest to both the region's early residents and today's visitors, including the Ochre Beds and Radium Hot Springs. Located in the northern half of the park, near where Ochre Creek flows into the aquamarine Vermilion River, the Ochre Beds are fed by iron-rich mineral springs. These "Paint Pots" drew not only the Ktunaxa, but likely even the Siksika and Pikuni of the plains. Though both much preferred their bountiful, open grasslands to the dark and daunting mountains, ochre was much sought after. Dried and packed into cakes, then pulverized and mixed with grease, it was used for decorating everything from clothing and tipis

Getting There: Kootenay National Park lies west of Banff and south of Yoho National Park. It is accessed by the Kootenay Parkway (Hwy 93), which links the Trans-Canada north of Banff townsite and Hwy 95. Crossing the Rockies at Vermilion Pass, the 94-km long parkway follows the Vermilion and Kootenay Valleys south and west. Most of the park's many interesting features can be accessed from the parkway; some, including Vermilion Pass, the Ochre Beds and Radium Hot Springs, are wheelchair accessible. A short trail to Dog Lake originates at the McLeod Meadows campsite, but many other, more challenging trails can be accessed from the parkway throughout the length of the park. Visitor centres in Radium Hot Springs and at Vermilion Crossing in the north half of the park provide information to travellers; more information can be obtained by emailing kootenay.info@pc.gc.ca

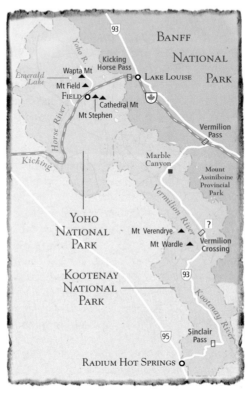

93
BANFF
NATIONAL
Yoho R.
Kicking
Horse Pass
Wapta Mt
Emerald Lake
Mt Field
FIELD
Cathedral Mt
Mt Stephen
Lake Louise
PARK
Horse River
Vermilion Pass
Kicking
Marble Canyon
Mount Assiniboine Provincial Park
Vermilion River
YOHO
NATIONAL
PARK
Mt Verendrye
Mt Wardle
Vermilion Crossing
KOOTENAY
NATIONAL
PARK
93
Kootenay River
95
Sinclair Pass
RADIUM HOT SPRINGS

Koalanuk

At THE NORTH END
of Windermere Lake, where
the Columbia River begins
its long journey to the sea in
earnest, is a spawning site
that drew the Tobacco Plains
Ktunaxa for at least a thousand
years. At the outlet of the lake, the
fledgling Columbia River narrows
for a short distance and then, where
Toby Creek flows in from the west,
the river widens to form a series of
shallow pools.

Archaeologists call this large fishing station the Salmon Beds Site, but
the Ktunaxa name for it—Koalanuk,
"where the lake empties into the river"
—is both lyrical and more descrip-
tive. Beginning in August and continuing into October, chinook salmon
once spawned here in large numbers
at the end of a remarkable 1900-
kilometre journey from the Pacific.

When the runs were good, the
Ktunaxa began the season by taking
salmon farther downstream, near
Briscoe and Fairmont Hot Springs,
moving up to Koalanuk as the season
progressed, in September or early
October.

Excavations at the site showed
that this was not simply a fishing
station. In fact, fishing may have been
a sideline. Here, stretching along the
west bank of the river for more than
a half-kilometre is fire-broken rock,
stone tools of Top-of-the-World
chert (see page 76), and hundreds of
animal bones, including bison, white-
tailed deer, elk, black bear and beaver.
The presence of large bison bones
(including a scapula or shoulder
bone) here, as well as at other sites

along the Rocky
Mountain Trench
make it clear that,
though the Ktunaxa were known to cross the mountains to hunt them in the eastern foothills, bison were
also found on the west side of the Rockies until just
prior to the settlement period. Many of the bones
were smashed, an indication that not only had the
meat been used, but the bones had been processed
as well, likely boiled for the fat they contained.

Chinook salmon were likely speared in the
shallows, for bone points of the type used in a leister,
or three-pronged fish spear, were recovered.

The Ktunaxa were not the only people to make
use of this food processing site. For the past two
centuries or perhaps longer, Shuswap from the
northwest—the Kinbasket people—also gathered
here in the fall. And ultimately, they stayed. Today,
the Kinbasket reserve occupies the east side of the
river north of Lake Windermere. Having intermarried
extensively with their Ktunaxa neighbors, the Kinbasket
people consider themselves to be part of the larger
Ktunaxa community, something that is made evident
by the official name taken by the original people of the
Kootenay region—the Ktunaxa/ Kinbasket First Nation.

Archaeologists have found other sites up and
down the Columbia Valley, and on terraces above the
Kootenay River to the south, which show that people
have been living in this bountiful place for at least the
past 5,000 or 6,000 years. On a high terrace overlooking the Kootenay River, excavations yielded a projectile point of black siltstone with broad deep corner
notches and a straight base. A similar point, found
near Kamloops, was believed to be between 4,500
and 6,000 years old and showed that people were still
primarily hunters of deer and elk, with their diets
supplemented by small game, birds and plants, as
well as salmon and other fish. Similar points found on

Leisters or three-pronged fish
spears, were also widely used
along the lower Fraser River
and in many places on the
West Coast. Finding them here
indicates that technologies
were widely shared.

Illustrations: Melanie Froese

To the west and south, the appearance of side-notched points, such as the Bitterroot point (top), signalled a major change in hunting technology, from spears to deadlier atlatls and darts.

The destinctive Pelican Lake point (bottom) was widely used across the west as far east as the plains of Saskatchewan.

The Ktunaxa roamed every part of the Kootenay River, from the glacial waters of its upper reaches, shown here, to its confluence with the Columbia.

Peter St. John

the Alberta plains mark the transition from spears to atlatls (or spear-throwers) and darts, a technological change that allowed hunters a much greater range.

On another high terrace above Windermere Lake, archaeologists found other indications that technologies were widely shared throughout the history of the Columbia Valley. Dated to between 2,100 and 2,400 BP, the site was a temporary camp used for processing game and fur-bearing animals, mainly elk, deer, wolverine, lynx and beaver. The two complete projectile points recovered were corner notched with expanding stems, very like Pelican Lake points found across Alberta and into Saskatchewan beginning about

3,300 BP. Yet it was not the atlatl points themselves, but rather the design that had been imported; the points and knives at the site were largely made of Top of the World chert and pale cream chalcedony.

Getting There: Koalanuk has not been marked or restored in any way, but can be reached by turning west on Toby Creek Road toward Invermere from Hwy 93/95 at the north end of Windermere Lake. Cross the Columbia River and turn north on Wilmer Road for just under two km to where it crosses Toby Creek. Park by the edge of the road and walk east to the river.

Moyie Lake, right, south of Cranbrook, is a kettle lake, formed by the melting of a huge block of ice, left in the wake of the retreating glaciers.

The Ktunaxa

THE KTUNAXA could justifiably claim to be British Columbia's first people. Their roots in the province go back more than 11,000 years, according to archaeologist Wayne Choquette, who has worked for more than three decades in the Kootenays. His work and that of others make it clear that the Ktunaxa (also known as the Kutenai, Kootenay and Kootenai) have inhabited the Columbia Basin without interruption since the last glaciation. From east to west, this huge area includes the western Crowsnest Pass, the upper Columbia and Elk Valleys, the Tobacco Plains, the entire

THE KTUNAXA

Revelstoke ○
Kootenay River
○ Golden
○ Calgary
BRITISH COLUMBIA
A L B E R T A
▲
▲
○ Cranbrook
▲
▲
▲

□ Traditional Territory
■ Current Territory
▲ Reserves

WASHINGTON | IDAHO | MONTANA

Kootenay Valley in British Columbia, northwestern Montana and the Idaho Panhandle, as well as the middle Columbia Valley south of Revelstoke,

Peter St. John

and part of eastern Washington. In addition, both Ktunaxa oral history and early fur trade history record the existence of an eastern group of Ktunaxa, who lived along the foothills of the Rockies in what is now Alberta. These buffalo hunters, who may have moved permanently out onto the plains when bison were largely extirpated on the western side of the Crowsnest Pass, apparently all died of disease, likely smallpox, soon after their first contact with Europeans.

Ktunaxa (pronounced Tun-ah-ha, with the "K" elided, the "x" pronounced as it is in the Greek alphabet, like the "ch" in "Bach", and the emphasis on the penultimate syllable) society might also be taken to be an original model for modern Canadian society as a whole, for it has long been egalitarian and remarkably forward thinking. Communal effort has always been valued and women are treated as fundamentally important to the well-being of the society as a whole. Female elders, for example, are believed to hold a great wealth of knowledge.

Ktunaxa territory is routinely described with superlatives; it has been called "one of the world's greatest game preserves" as well as "Canada's most stunning mountain scenery". With the exception of the seashore and the desert, it includes almost every natural environment found in Western Canada: the vast wetlands of the Columbia Valley; the hot, dry grasslands of the Tobacco Plains; the rich valleys and deep lakes of the Kootenay River; the forested slopes, craggy uplands and glaciated heights of the Rockies and Columbia Mountains and, until the twentieth century with its many dams, two of the most bountiful rivers on the continent.

Moreover, though the valleys of the Kootenay Rockies are natural travel corridors, which is likely how the Ktunaxa found their paradise so many millennia ago, thanks to the steep and craggy mountain ranges to the north and east, until quite recently, few found their way into this remarkable corner of B.C.

In all the thousands of years before, the Ktunaxa way of life was simple, generally portable and attuned to the land.

The earliest inhabitants arrived even before the glacial lakes, which once almost filled the narrow valleys, had drained. Their hunting camps have been found on south and southwest-facing slopes of tributaries of the lower Kootenay River, including the Goat and Moyie Rivers.

Archaeologist Choquette has identified what he believes are traces of two early cultures in the upper Columbia and lower Kootenay basins. The earliest were found in a series of workshops high on ancient lake terraces and beaches above the Moyie River in the southwest Purcell Mountains. Here, he found large sidescrapers, flaked tools and spear points—some more than ten centimetres long—quarried of local black tourmalinite, which can be flaked to a very sharp edge. Excavated a full metre below a layer of ash from the eruption of Mount Mazama 7,640 years ago, the tools and weapons were likely made in one of a series of summer hunting and quarrying camps along the Great Terrace, which was formed by Glacial Lake Columbia on the sides of the Columbia and lower Kootenay Rivers between 12,000 and 11,500 years ago.

Choquette believes the Moyie River campsite is more than 11,000 years old and that these early residents of the Kootenays came from the south, for their large, often reworked, biface tools and shouldered or lanceolate spear points, though quarried of local stone, as well as their lakeside camps, were similar to the tools, weapons and camps of early Americans who lived in the Great Basin at the end of the last glaciation. Traces of this "Goatfell Complex", as he calls it, have also been found in Montana and Idaho.

The origins of the Ktunaxa are still being debated and their language is termed a "linguistic isolate". It is possible, however, that they may be part of the Uto-Aztecan linguistic tradition, which includes families of languages spoken in various parts of the Southwest and throughout much of Mexico and into Central America.

Recent Canadian studies have suggested that Ktunaxa and Proto-Salish, the common ancestor of all the Salish languages, had an ancestor language in common, one that linguists have dubbed Proto-Ktunaxa-Salish, which not only predates Proto-Salish, but indicates an ancient connection between Ktunaxa and Salish.

The word "Kootenay", may be derived from the Ktunaxa *quthni*, which means to "travel by water".

Eleven thousand years ago, the Great Terrace, above the Moyie River Valley, was waterfront property. The lovely lanceolate spear point (above) is on display at the Creston Museum.

Dennis Fast, courtesy of the Creston Museum

Peter St. John

And the earliest Ktunaxa lived by the shores of the great glacial lakes. The people of this early culture were mainly big game hunters, a lifestyle determined by their environment. Like a blank canvas, in the earliest post-glacial period, British Columbia's icy glacial lakes and rivers were largely empty of fish. The Columbia River was the exception to this norm. With its mouth south of the ice sheets, it is clear from both the total number of native species of fish that populate the river today and the proportion of species that are salt-water intolerant that it served as a refuge for freshwater fish during the long glaciation.

Naturalist E.C. Pielou writes that the Columbia River system has a total of forty-five native species, compared with thirty-nine for the Fraser River and just twenty-seven for the Stikine and Nass River systems in northern B.C. Moreover, more than half the Columbia River species are salt-water intolerant, compared to less than twenty per cent in the two northern river systems. Even so, it would have taken the fish time to repopulate the upper reaches of the Columbia River, to reach Lake Kootenay in significant numbers, for example. Also, in the earliest

post-glacial period, the falls where salmon and other fish might easily be taken at spawning time were drowned beneath glacial meltwater. As a result, catching them would have required boats, which may have been in short supply.

As for their long history in the region, the Ktunaxa, and their American relatives, the Kootenai of Idaho and Montana, will tell you that they have been in the region since the beginning of time and more than their language makes it clear that they've lived, largely isolated, in the Columbia Basin for thousands of years. Certainly their extraordinary sturgeon-nose canoes (see page 90), which are perfect for the region's narrow lakes, rivers and marshlands, are unlike any others in North America. In his excellent book, *The Algonquin Birchbark Canoe*, author David Gidmark writes that in "only one area, that of the Amur River in southeastern Siberia, were Birchbark canoes of any importance, and here is found the only prototype for an American canoe."

Beginning in the nineteenth century, others also pointed to the striking similarlities between the Ktunaxa canoe and the monitor-shaped canoes of the peoples of Siberia's Amur River region. It's a small step to consider that the technology needed to create such magnificent craft first arrived in North America with early Asian immigrants.

In addition to Ktunaxa sites, archaeologists have found what appear to be the remains of another, more recent culture near the confluence of the Kootenay and Columbia Rivers. From artifacts found at two sites, one near the uppermost falls on the western arm of Kootenay

The St. Mary River, one of British Columbia's premier trout streams, tumbles over the rocks north of Cranbook. Sites excavated along the river date back more than 3,000 years.

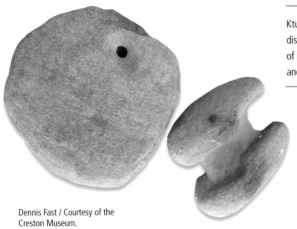

Ktunaxa canoe anchors, such as these on display at the Creston Museum, were part of a lifestyle supremely adapted to the rivers and lakes of the region.

Dennis Fast / Courtesy of the
Creston Museum.

Lake and the other on a high terrace above the confluence, it is clear that these people had very different tools and weapons than did their earlier neighbours in the Moyie River Valley. Using microblades of argillite and rhyolite similar to those found in the province's far northwest, as well as cobble sinkers, it seems that they were fishermen. Moreover, these people made bone into tools, much as coastal cultures did. Archaeologists therefore speculate that these were northern people who may have reached the Kootenay region by the same route the salmon did, by following the Columbia River upstream. Were these the ancestors of the Sinixt people, an Interior Salish culture that inhabited the West Kootenays for thousands of years, only to be devastated by smallpox in the nineteenth century? And were they related to the Ainu, the ancestral people of Japan, whom it increasingly appears may have followed the shoreline of the northern Pacific and Beringia to North America at the end of the last glaciation? Forensic studies of Kennewick Man, whose magnificently preserved 9,400-old-skeleton was found along the lower Columbia River between Washington and Oregon, have shown that his closest relatives were Asian, and that his diet was mainly marine aquatic life.

Many questions are yet to be answered; indeed, as Choquette has stated time and again, the Kootenay region has been neglected archaeologically and much is still to be learned. Despite that, signs of the long sojourn of the Ktunaxa are everywhere, from Top-of-the-World (see page 76), a plateau located 2600 metres up in the Hughes Range east of Skookumchuk, to the Salmon Beds Site (see page 65) just downstream from the headwaters of the Columbia.

The Ktunaxa themselves are still very much present, both in the many communities in the area and on six reserves in the region: near Windermere, close to the headwaters of the Columbia River; on the St. Mary's River near Cranbrook; at Grassmere on the Tobacco Plains; at Creston on the Lower Kootenay; on the Flathead Reserve in Montana and at Bonner's Ferry in the Idaho Panhandle. Together, they make up six of seven members of the Ktunaxa/ Kinbasket Tribal Council; the seventh, the Kinbasket, are based at Althamer, near Invermere, B.C., and represent the descendants of the Shuswap that moved into the region about 200 years ago.

Much more about Ktunaxa history and culture, as well as their painful relationship with settlers in the region during the past century, can be learned at the Yaqan Nukiy Heritage Centre at 3560 Highway 21 in Creston and at the Ktunaxa Interpretive Centre in the newly renovated and rehabilitated St. Eugene Mission Residential School just outside Cranbrook. It says much about the Ktunaxa themselves that they have turned what was a place of such trauma and loss into a place of healing and accomplishment.

Ktunaxa Interpretive Centre
at St. Eugene Mission Resort

IT WAS MARY PAUL, a respected Ktunaxa elder, who spelled out the vision that guided her people in the transformation of a place of pain into a place of opportunity and healing.

The residential school at St. Eugene Mission was where the culture of the Ktunaxa had been taken away. "Now", she said as the Ktunaxa considered what might be done with the site, "it should be within that building that it's returned."

Today the transformation is complete. Dark halls once filled with homesick children now bustle with staff and visitors. A beautifully appointed 125-room hotel, luxurious spa and a casino blend heritage architecture with superb accoutrements. And to ensure that the culture they nearly lost is not only revived, but shared, the resort includes an interpretive centre with museum quality exhibits and an arts and crafts co-op. Outside, a tipi village beckons and one of the best new golf courses in Canada winds through rolling woodlands along the sparkling St. Mary River.

The river has been central to the Aqamnik, the "people of the thick woods", for millennia. One of British Columbia's premier trout rivers, it originates in the Purcells and tumbles east through St. Mary Lake to join the Kootenay River just north of Fort Steele. Even today (thanks to modern fishing regulations) from April through October, the crystalline water of the river is alive with brook and bull trout, cutthroat, dolly varden and burbot or freshwater ling cod. All this was even truer in the past, for in addition to trout, in dominant years the river teemed with spawning salmon from late June to October. One of the most popular sites, particularly in September when the kokanee spawn, was the confluence of Joseph Creek. Archaeologists have found a campsite here that dates back 3,000 years.

Getting There: From Hwy 3 just northeast of Cranbrook, take Hwy 95A northwest toward Kimberley. Travel for one km, then turn north onto Mission Road. St. Eugene Golf Resort and the Ktunaxa Interpretive Centre are located at 7468 Mission Road, five kilometres from the junction with Hwy 95A. Check online for interpretive centre hours.

Photos: Peter St. John

The Kootenay River

Dennis Fast

Grizzly mothers and cubs can be found in many places along the Kootenay River.

THE KOOTENAY RIVER originates high in the Beaverfoot Range of the Western Rockies and, fed by dozens of creeks and several rivers, rolls southeast through Kootenay National Park and then west to Canal Flats. Separated from the headwaters of the Columbia River, which it will eventually join, by less than two kilometres, it turns south instead and heads down the Rocky Mountain Trench into the United States.

But the call of Canada is strong and east of Libby, Montana (where today the huge Libby Dam impounds the river's water, creating Lake Koocanusa—the name is a combination of "Kootenay, Canada and USA"—the river turns northwest, crossing the Idaho Panhandle and the international boundary and reentering B.C. just south of Creston. Here, the current slows to create bountiful wetlands in the wide valley (see page 87), before flowing into Kootenay Lake in the Purcell Trench, a deep cleft between the Purcell and Selkirk Mountains.

Today, Kootenay Lake is 107 kilometres long along its main north-south axis, with a west arm, through which the lake empties into the Columbia River, of forty-five kilometres. Fjord-like, it averages less than four kilometres in width, has an average depth of ninety-four metres or 300 feet and stands at an altitude of 532 metres above sea level (asl), varying by only about three metres in the spring.

For nearly 1,000 years following the last glaciation, however, it was much longer and deeper. Geologists have found the margins of Glacial Lake Kootenai at 762 metres asl—fully 230 metres above the shores of today's lake. Below those ancient water lines are two incontestable date markers, the ash from the eruptions of Mount St. Helens 11,800 years ago, and Glacier Peak, a little-known, explosively active volcano that erupted in west-central Washington about the same time.

After Lake Kootenai drained, a second glacial lake, dubbed Glacial Lake Columbia, filled the Purcell Trench, stretching upstream along the Kootenay River to east of Libby and downstream, past the confluence of the Kootenay and Columbia Rivers, to fill the Columbia River Basin from Revelstoke south to Grand Coulee, Montana. Glacial Lake Columbia left as its indisputable imprint a shoreline that geologists call the "Great Terrace" at 595 metres asl, almost sixty metres above today's lake. These two glacial lakes provided a conduit to both Kootenay Lake and the Okanagan Valley for kokanee salmon (see page 121), which would, over time, forever abandon their anadromous or ocean-going tendencies, becoming freshwater fish throughout their lives.

Lake Koocanusa, seen here at Tobacco Plains, stretches across the U.S. border.

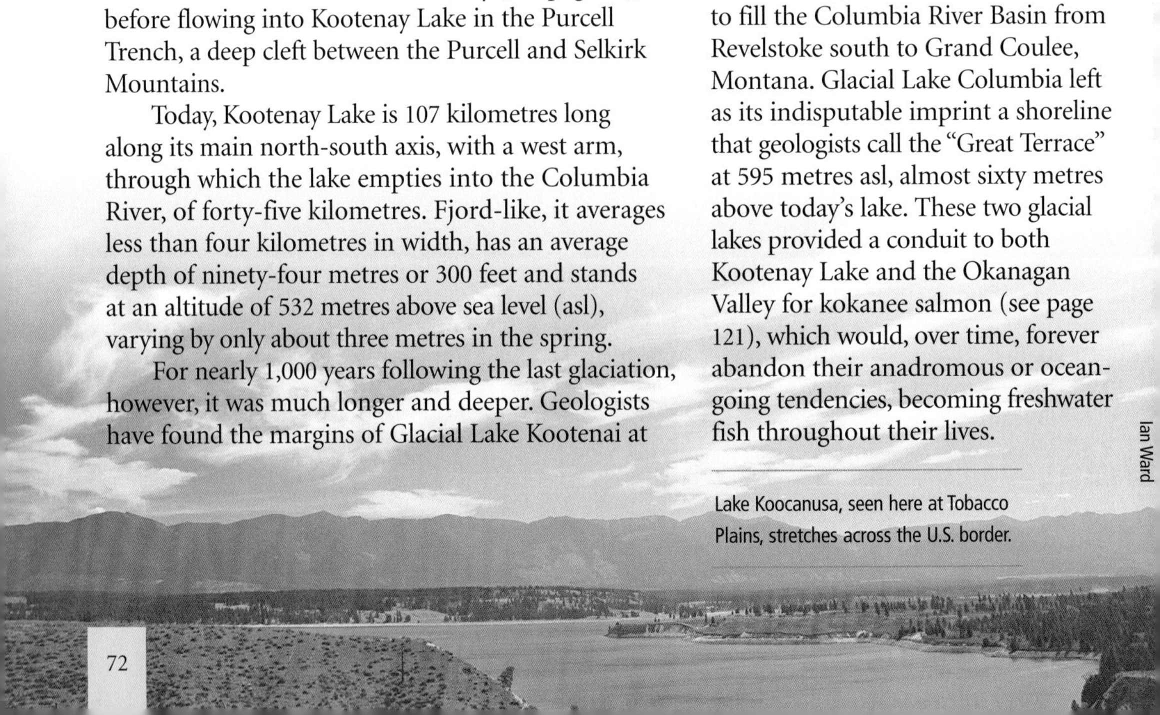

Ian Ward

Salmon are not the only sought-after inhabitants of Kootenay Lake. The lake and its tributaries to the north are also famous for their huge trout.

Bonnington Falls (below) are now dammed, but once created a barrier to sockeye salmon.

Though the waters of these long, ribbonlike glacial lakes were undoubtedly cold, the slopes and terraces above, particularly on the south and west-facing sides of the valleys, were habitable soon after the ice melted, thanks to two factors. The first was the global climate, which was much warmer than it is today; it was, of course, this remarkable global warming that led to the worldwide melting of the enormous ice sheets. The second was the "solar bowl", as archaeologist Wayne Choquette has called it, formed by the way the "south end of the Monashee and Selkirk Mountains … meet the Purcell-Cabinet-Bitterroot chain to form a great arc bounding [an] enclosed basin on the north and east."

With a high pressure centre over the huge Laurentian ice dome to the east, the easterly flow of air would have drawn warmer air from the southwest into this solar bowl, creating a micro-climate of the kind experienced today by Calgarians on a regular basis. These "chinook-like" conditions created an environment where spruce and fir, sage and grasses thrived. And these, in turn, drew large mammals and, hard on their heels, human hunters.

By 9,500 BP, the glacial meltwaters had drained, and Glacial Lake Columbia was no longer. Lower Bonnington Falls, between the modern cities of Nelson and Castlegar, became the rocky valve that would control the entire Kootenay River and Lake system.

Bonnington Falls spelled the end of anadromous sockeye spawning in Kootenay Lake. Every fish species has an upstream "burst speed", a rate of maximum swimming energy output it can maintain for no more than a few seconds before rest is required. The onrush of water at Bonnington Falls exceeded the burst speed abilities of the anadromous sockeye, and thus became a permanent biological barrier. Sockeye spawners were forced to turn back, but in the vast water system above the falls, the kokanee salmon of Kootenay Lake and the Kootenay River had already nicely filled the ecological gap. Though similar in appearance and technically still able to mate and produce offspring, the sockeye and kokanee had become separate species.

NINGTON FALLS KOOTENAY RIVER.B.C.

White Sturgeon

VIRTUALLY UNCHANGED over the past seventy million years and with ancestors that date back 200 million years to the early Jurassic period, sturgeon are among the oldest fish in existence. Little wonder they have been called living fossils. Unlike most other, more "modern" species, only parts of a sturgeon's skeleton are bony; the skull, for example, is made of cartilage, as are most of the vertebrae. With diamond-shaped plates running along the sides of their bodies, white sturgeon (which are in fact gray; it is their flesh that is white) have also been called "diamond-sides". In optimum circumstances, anadromous white sturgeon (those that migrate between fresh and salt water) can reach nearly six metres in length and weigh more than 800 kilograms. Members of the species that are confined to fresh water are considerably smaller.

Found only in the Northern Hemisphere, but once widely distributed, almost all the world's twenty-four and North America's eight species of sturgeon are threatened or endangered. In large part, this is due to two conflicting realities: the very low reproductive rate of this ancient species and the highly destructive capacity of human technology.

Depending on the species, male sturgeon, which mature sexually earlier than females, are able to reproduce at age sixteen, but female sturgeon may not mate and begin to spawn until they are between twenty-two and thirty years of age. And spawning, which may only occur every six years, requires very specific conditions to produce viable young. White sturgeon, for example, require swift-flowing streams with pebbly or cobbled beds for successful spawning.

Yet these ancient creatures proved able to adapt to changing conditions in the past. When Bonnington Falls, just upstream of the confluence of the Kootenay and Columbia Rivers, became impassable as a result of isostatic rebound—the rising of the earth's surface after the enormous sheets of ice melted—and the draining of the glacial lakes, white sturgeon both upstream and downstream of the falls were able to adjust to the new reality. Those in the Columbia River system downstream of the falls found new spawning channels and continued their anadromous existence that took them to the Pacific and back, while it seems that at least some of those in the Kootenay River system above the falls had already given up the sea forever and adapted to a purely freshwater existence.

Both sturgeon populations had very long life spans, with individuals living 100 years or more. And for thousands of years, the people of British Columbia lived in harmony with these ancient creatures, taking no more than the species' low reproduction rates could sustain. But the human-engineered changes of the past century have proven deadly. Overfishing, dam building that has altered the flow and depth of nearly every major river in British Columbia, destruction of marshes and sloughs that have long served as nurseries for young fish, as well as pollution, have combined to put these relics of the dinosaur age on the path to extinction.

Though fishing for sturgeon in the Kootenay system has been illegal in B.C. since 1990 and was outlawed in Montana and Idaho prior to that, the landlocked white sturgeon population is in steep decline. Confined to the Kootenay Lake and about 270 kilometres of the Kootenay (or Kootenai, as it's spelled in the U.S.) River between the lake and Kootenai Falls in Montana, these remarkable fish have been dramatically affected by the Libby Dam, which began

Peter St. John

operation in 1974. Disrupting the natural flow, altering daily and seasonal water temperatures and burying the crucial pebbled spawning sites in silt and sand, the dam has made reproduction in the river all but impossible. As a result, the

with them, in 1998 British Columbia Environment approved the use of the Kootenay Trout Hatchery near Fort Steele as a backup or fail-safe facility for fertilized eggs and juvenile sturgeon, to ensure that at least some fish would survive should a catastrophe occur at the Idaho hatchery. More recently, the hatchery has begun raising sturgeon young to be released into the upper Columbia River.

This baby sturgeon makes it hard to believe, but the largest individual on record in the landlocked Kootenay River population was a 159-kilogram (or 350-pound) fish captured in Kootenay Lake in 1995. It was estimated to be between eighty and ninety years of age.

Dennis Fast

total population had declined to approximately 1,468 wild fish by 1997, according to the U.S. Fish and Wildlife Service. And of the remaining sturgeon, almost none were under the age of twenty-five. At an attrition rate of approximately nine per cent, researchers predict that all remaining wild fish will be gone by 2065 and that the wild population will be functionally extinct by 2035.

In an effort to save the fish that have meant so much to their people for so many millennia, the Kootenai of Idaho have begun to raise young sturgeon in a hatchery in Bonners Ferry, releasing them annually into the river. Working in partnership

Meanwhile, the engineers at Libby Dam continue to experiment with flow capacity, in the hope that some solution to the spawning problem can be found without disrupting power production or flood control.

Getting There: The chances of seeing a white sturgeon in the wild are relatively remote, but the Kootenay Trout Hatchery has excellent visitor facilities, including aquariums, educational models and displays, and offers self-guided tours, an outdoor turtle pond and beautifully groomed hatchery grounds where visitors are encouraged to picnic. The facility, which is wheelchair accessible, is open year round from 8 a.m. to 4 p.m. Admission is free. From Cranbrook, travel 32 km southeast on Hwy 3 to the Wardner/Fort Steele Road. Go north six kilometres, following the signs.

Top-of-the-World Provincial Park

CLUES TO THE PAST can be found in the most unlikely places. This is true of Top-of-the-World Provincial Park. Here, high in the western ranges of the Rockies northeast of Kimberley, are some of Canada's best-preserved examples of the world's distant maritime past.

It is the last of these, the stromato-poroids, that are particularly striking here, for these spongelike organisms grew in tall colonies that may, in life, have been as much as five metres high—the tallest organic

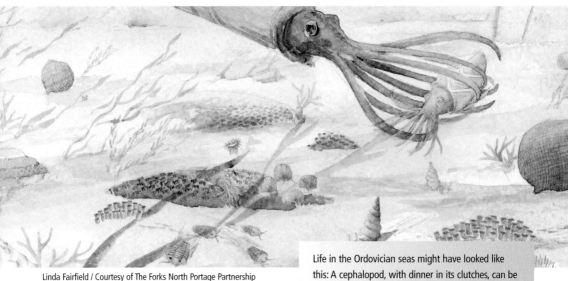

Linda Fairfield / Courtesy of The Forks North Portage Partnership

Life in the Ordovician seas might have looked like this: A cephalopod, with dinner in its clutches, can be seen above chain corals. At the right of the painting is a receptaculitid, the balloon-like plant growing out of the sea floor. Similar fossils are found in Manitoba, for a shallow seaway inundated much of central North America during the Ordovician.

About 450 million years ago, long before terrestrial life on Earth evolved, what is now eastern B.C. lay at the edge of Laurentia, the nucleus of ancient North America. Sea levels were unusually high during this period and as Laurentia floated astride the equator, much of it was covered with shallow seas. This warm, clear water proved a perfect incubator for life.

Among the great variety of bottom-dwelling organisms that thrived in the Ordovician seas were hard-shelled invertebrates: chain corals, receptaculitids (an extinct form of algae that lived in large colonies), large straight cephalopods (distant relatives of the pearly nautilus) and early forms of stromatoporoids.

constructions in the Ordovician seas.

When all these creatures died, their bodies littered the floor of the sea, gradually building layer upon layer of what would eventually be limestone littered with fossils. Geologists call this "calcareous biogenic debris"; "calcareous" refers to the

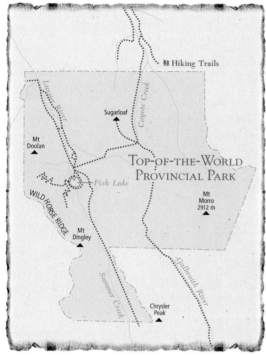

Ironically, despite the tropical beginnings of these primitive animals, fossils of this age and type are known to geologists as "Arctic Ordovician Fauna". The "arctic" label was given because they were first identified in rocks in the Canadian Arctic and Greenland. Like Top-of-the-World park, these sites not only seem unlikely today, but are testament to the power and ultimate idiosyncrasy of plate tectonics.

The outcrops at Top-of-the-World park have long been known to the Ktunaxa, who have made southeastern B.C. their home for more than 10,000 years. For the past several thousand years, they made regular trips to Top-of-the-World to hunt, and to quarry slabs of dark gray chert from the limestone outcrops. This stone was not only used to create weapons and tools that were harder than steel, but to trade to the Shuswap to the north and the Kootenai and Shoshone to the south.

Most of the park is above 1,800 metres (or 5,940 feet), high in the Kootenay Range of the Rocky Mountains and is noted for its beauty as well as its paleontological and archaeological wonders.

Getting There: While there are no roads through Top-of-the-World Provincial Park, its trails can be accessed from Sheep Creek Road off Hwy 93/95, just south of Skookumchuk. Though rough and not recommended for low-clearance vehicles, it is well marked. Since this route is a logging road; use caution at all times. It is normally passable from late May to mid-November. Back-country camping is permitted at Fish Lake, where there is also an overnight cabin that sleeps 20 on a first-come basis, and Coyote Creek. Water is almost non-existent in some parts of the park and no supplies are available, so campers need to be fully self-sufficient.

calcite—the mix of calcium, carbon and oxygen—that makes up the shells, and "biogenic" means that the origin of the debris is living organisms—the body parts of the creatures in the sea.

The fossils are particularly impressive in the limestones at Top-of-the-World, because the original calcite has been replaced by silica, or silicon dioxide, which has the same composition as chert, a stone that is particularly hard and resistant to weathering. As a result, the fossils stand conspicuously above the limestone, which dissolves more readily over time. The stromatoporoid fossils, which are up to two metres long, can be easily identified because they look like weathered fence posts.

Snow-clad mountains reflected in Duck Lake, just south of Kootenay Lake in the Creston Valley wetlands, herald the coming of spring and the arrival of great flocks of migrating waterfowl.

Dennis Fast

The Crowsnest Highway

Picturesque and historic,

the Crowsnest Highway (No. 3)
is British Columbia's most southerly route
from the Rocky Mountains to the Fraser River.

FOR MOST OF ITS LENGTH, it hugs the American border, all but touching it in places such as Cascade in the Monashee Mountains and Osoyoos in the South Okanagan. Throughout, it is remarkably scenic and increasingly favored by skiers and snowboarders, as well as birders and bikers. But nowhere is it a road to be rushed.

Following it from east to west, the way B.C. was built (though the road, or trail, itself was constructed in the other direction, beginning in 1859), the Crowsnest Highway begins on the Alberta border at Crowsnest Pass and carries its name for 590 kilometres west to Hope.

Joining the Elk Valley (see page 81) at Sparwood, it follows the Elk River past the ski slopes of Fernie and what are likely the world's oldest cottonwood trees (see page 84) to the Rocky Mountain Trench (on page 48). Here, the landscape and climate both abruptly change. From the narrow, densely forested Elk Valley, with its deep snowfall and bountiful rains, travellers find themselves in open grasslands, often under sunny skies. To the south are the Tobacco Plains (page 85) where, until less than 200 years ago, bison roamed. To the north is the Columbia Valley (see page 55), which cradles both the upper Columbia and Kootenay Rivers.

Crossing the Kootenay just above Lake Koocanusa (named for the river and the countries it joins), the highway heads north to join Highway 93/95 just north of Cranbrook. Ten kilometres north, is Fort Steele, one of a chain of Northwest Mounted Police forts that were erected, like beads on a string, in the 1880s to keep the peace during the settlement period. Today, the fort is a heritage town, where history is reenacted during the summer months. But its presence here recalls the other name for the Crowsnest Highway: the Red Coat Trail.

From the junction of Highway 93/95, No. 3 turns south to follow the Moyie River through the Purcells. This is one of the oldest inhabited regions of Canada, boasting archaeological sites on some of the valley's

Dennis Fast

upper terraces that date back to the waning days of the last glaciation.

Then, just ahead, is Creston (see page 87), centred on the wide valley and opulent marshes of the lower Kootenay River. This is a birders' paradise, with nearly 300 species recorded, including some that breed nowhere else in B.C. It is also home to dozens of mammals, including the gravely endangered mountain caribou (on page 91), as well as the only place in the province where leopard frogs are still found. Thanks to the efforts of many local groups as well as the provincial government, the valley's irreplaceable wetlands and mountain forests are being increasingly protected.

West of Creston, the Crowsnest Highway winds through the southern Monashees, dipping into the stunningly picturesque Columbia Valley at Castlegar and climbing past Nancy Green Provincial Park, which draws both alpine and nordic skiers, to Bonanza Pass. Skirting the U.S. border at Grand Forks and Midway, the highway traverses the Okanagan Highlands (page 112), before descending into the South Okanagan and the centre of Canada's only real desert (see page 124).

Following the valley to Princeton, the Crowsnest crosses the Similkameen River. Now, just 128 kilometres separates travellers from the Fraser Valley. Yet this is one of the province's most notorious sections of road. And though it's greatly improved over the last two or three decades, as it passes through the Cascade

Above: Mount Moyie is mirrored in the still water of Moyie Lake.

Mountains, there are reminders of the challenges of the past.

In January 1965, the southwestern slope of Johnson Peak collapsed, spreading nearly fifty million cubic metres of rock and mud over a three-kilometre stretch of the infamous Hope-Princeton Highway. Though the landslide took place in midwinter and in the early morning hours, four people were nevertheless killed. Other slides in this geologically unstable region have also been disastrous (see pages 150 and 161).

Hope marks the end of the Crowsnest Highway (a road that actually begins in Manitoba), as it joins the Trans-Canada here, but there is much more for travellers to see, whether they are heading north to the Fraser Canyon (on page 133) or west along the Fraser Valley (page 157).

The Elk and Flathead Valleys

NORTH OF ITS CONFLUENCE with the Kootenay River, the Elk Valley is narrow and guarded by snow-capped mountains. Centred on the braided current of the Elk River, which twists around pebbled, willow-crowned islands, it's a picturesque, if often damp or snowy place. Locals love it for its increasingly renowned ski slopes, its relaxed

atmosphere, internationally important coal fields and world record cottonwoods (see page 84).

North of Fernie, the Elk Valley is an extension of the Flathead Valley to the southeast, a remarkable half-graben (pronounced half grah-ben) or block fault in which the rock on the west side of the valley has dropped almost straight down, relative to the rock on the east. As Ben Gadd

Photos: Ian Ward

explains in the *Handbook of the Rockies*, it's "like a long, skinny door opening downward".

This door into the earth is an almost unbelievable six kilometres deep, more than twice as deep as the surrounding mountains are high. Thanks to the work of rivers and glaciers, much of that depth is filled with sediment. However, as the floor of the trench has sunk, its eastern wall has risen, exposing the layers and providing tantalizing evidence of the region's ancient beginnings.

About 150 million years ago, as dinosaurs were beginning to flourish in British Columbia, what is now the southeastern corner of the province lay along the lush, humid coast of Pangea, of which included ancestral North America. This coastal shore was enveloped by subtropical forests, swamps and wetlands. As the vegetation died, it was effectively sealed beneath sediments and covered by a narrow sea that lay east of the nascent mountain ranges growing along the continent's western edge. By the time the mountain-building spread eastward with the creation of the front ranges and foothills, the ancient peat had aged to become coal.

Some of this resource, which would one day prove precious to settlers in the valley, was forced deep into the earth by the folding and faulting, but in other places, it was exposed on the mountainsides by erosion, making it easy

Above and below: The bowl-shaped slopes of the Lizard Range above Fernie, which some have likened to an open catcher's mitt, are key to the region's winter popularity.

Centre: The ski hills just south of Fernie, minus the snow.

Ian Ward

The coal found on Morrissey Ridge, southeast of Fernie (above), and along Coal Creek, which runs just north of it, was responsible for the initial creation of the town.

for early Europeans to identify as they passed through the area. This was the case at what would become Fernie. In 1845, while travelling down the Elk Valley with the Ktunaxa, Belgian Jesuit missionary Pierre-Jean DeSmet found a large piece of coal. According to local historian John Kinnear, he mentioned the coal in a letter to his bishop in New York. Presciently, he added, "I am convinced that this fossil could be abundantly procured."

The outcroppings of coal were also noticed by later travellers, including renowned geologist George M. Dawson, who mapped the formations for the Geological Survey of Canada in 1886. Mining began eleven years later, and initially focused on the relatively flat, and thus easily mined, seams at Coal Creek, east of the present town of Fernie.

Later mines were both deeper and more hazardous. Still, the East Kootenay coal fields—those in the Elk and Flathead Valleys, along with the Crowsnest Pass Formations—constitute the major coal fields in B.C. and continue to contribute significantly to the regional economy. Yet, ever more aware of both the long-term environmental effects of mining and the

region's other, more sustainable attributes, residents of the region are reigning in the ambitions of mining companies as never before. Exploration for coal bed methane was recently abandoned by one of North America's largest petroleum companies, which found the environmental expectations too high.

There are other reminders of British Columbia's Jurassic past in the region, including fossils of clams and cephalopods; logs and stumps 145 million years old, and—perhaps most famously—a giant ammonite fossil that was found near Fernie.

Purportedly reported as resembling a "fossil truck tire" by a student in 1947, Canada's biggest ammonite was dubbed "*Titanites occidentalis*" by Hans Frebold of the GSC, though his colleagues later realized that the Fernie colossus, which measured almost 1.5 metres across, was not of the genus *Titanites*, a common type of large ammonite found in Jurassic rocks in Dorset, England. The name remains, however, for to date, the titan has never been properly identified.

The people of the Elk and Flathead Valleys—the Ktunaxa (see page 67)—have lived and hunted here since the retreat of the glaciers almost 11,000 years ago. At home anywhere in the region, they once spent their winters to the south, on the warm, sunny Tobacco Plains, and in the grasslands and wetlands of Creston, travelling north in the spring in their magnificent sturgeon-nosed canoes to fish for trout and char, and to wait for the

The confluence of the Elk and Kootenay Rivers south of Fernie is marked by the striking Elk Valley Hoodoos. Here, wind and water have carved the glacial sediments into weird and wonderful shapes.

spawning salmon and sturgeon in late summer.

Bison could be found in the grasslands of the Kootenays until about a century ago; there are reportedly two "buffalo jumps" in the valley, though their locations are carefully guarded. In the fall, however, the Columbia Valley Ktunaxa often traversed the Rockies over the Crowsnest Pass, which offered easy passage—relatively speaking—to the bountiful buffalo plains of what is now southern Alberta. Their encampments have been found on the high, dry, south-or southwest-facing slopes of the north side of the valley from Elko, B.C. to Lundbreck, Alberta, and north to Banff and beyond.

The mountain meadows and forested slopes were alive with deer, elk, mountain goats and sheep, but the Crowsnest Pass offered something more—access to Livingstone Quarry, where the hunters mined steel-hard chert for spear points, knives and scrapers. (For more on this, see the sections on the Crowsnest Pass and Livingstone Quarry in *In Search of Ancient Alberta*.)

In short, for millennia, for the people of the Elk and Flathead Valleys, life was bountiful in a land tucked away from the world.

Ian Ward

Getting There: Hwy No. 3 (the Crowsnest Highway or the Red Coat Trail as it is also known) runs along the Elk Valley from Sparwood to Elko. The Flathead Valley can be accessed from No. 3 via the Byron Creek Mine Road, which connects to the Flathead Forestry Road, a conduit to the picturesque Akamina-Kishenina Provincial Recreation Area, which many would like to see protected as an addition to Canada's Waterton National Park on the west and Glacier National Park in Montana to the south.

The Elk Valley Hoodoos can be found by following Hwy 93 less than 10 km south from the Crowsnest Highway to the mouth of the Elk River.

Morrissey Grove

ON THE EAST SIDE OF THE ELK RIVER, along a shoreline crowded by young cedars, stands a grove of cottonwood trees that may be among the oldest of their species on Earth. Among them are trees that were seedlings when Samuel Camplain founded the first settlement in Québec. By the time David Thompson made his epic journey to the Pacific Coast, most of the trees in the Morrissey Grove were more than thirty metres tall; a century later, as Donald Smith drove the last spike of the railway that joined Canada from sea to sea, all would have been considered old for the species. Yet even today, though topped and scarred, these venerable trees live on. Naturalists believe the eldest among them is four centuries old; until these were dated, the oldest known cottonwood was 250 years old.

Directed to the grove by residents of Fernie, University of Lethbridge biologists Stewart Rood and Mary Louise Polzin, studied and measured the huge trees as part of a binational study of all five *Populus* species that ranged from British Columbia's Peace River to California and Colorado.

Though we may never know whether these really are the world's oldest cottonwoods, the Elk Valley grove is remarkably ancient, with trees ranging in age from almost 250 to more than 400 years old. On average, this fast-growing riparian species lives less than 150 years. Its habitat, along streams and rivers that are subject to flooding and progressive meandering, means that many cottonwoods are undercut and toppled by the very rivers they depend on for water and life.

Despite heights between forty and forty-eight metres and trunks that range in circumference to almost ten metres, these are not the largest black cottonwoods on record. Larger trees have been found in coastal B.C., Oregon and Washington, but those trees proved to be younger, with larger annual growth rings.

Located on the east side of the Elk River

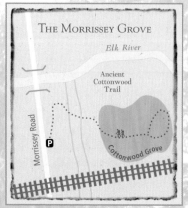

THE MORRISSEY GROVE

Elk River

Ancient
Cottonwood
Trail

Morrissey Road

P

Cottonwood Grove

south of Fernie, Morrissey Grove is easy to find, thanks to the efforts of the Nature Conservancy and Tembec, a forest products company, which recently built a signed trail among the trees. The presence of the grove tells scientists a number of things. First, situated less than 100 metres from the water, it provides proof that the Elk River, though prone to spring flooding, has stayed—at least along this stretch—within its banks for the past four centuries.

Second, it may be that the cool winters and comparatively short growing season on the western slopes of the Rockies are responsible for the very narrow annual growth rings found in the Morrissey Grove; the rings average just 2.7 millimetres or an eighth of an inch. Interestingly, younger trees at the same site have considerably larger annual growth rings, confirming that the climate has warmed significantly since the Little Ice Age, which prevailed when the venerable cottonwoods were young.

Though used when young in light construction, for tipi poles or building travois, and sometimes tapped for their sweet, light syrup, poplars were not prized by early British Columbians in the same way that cedar and lodgepole pine were. Still, this is a precious stand of trees.

Getting There: Follow Hwy 3 about 14 km south of Fernie to Morrissey Road. Turn east and cross the Elk River. Park in the small pullout area immediately east of the bridge. The trail, which is well signed, is immediately north of the road. Given the propensity of dead branches to fall from considerable heights, it is best not to visit during periods of high wind.

The Tobacco Plains

NORTH OF MORRISSEY CREEK, the Elk Valley is narrow and moist, a place where cedars thrive and snow comes early and lingers late. But south of this point, the Elk River suddenly emerges from its steep-sided, heavily forested valley into open forest and grasslands. Just south of Elko, the river turns west and pours into the Kootenay River between high sand cliffs. Hoodoos line the walls of the valley, showing how much the rivers have cut down since Glacial Lake Kootenai laid down its thick bed of sand and gravel about 12,000 years ago.

To the south, along Highway 93, are the Tobacco Plains. Irrigated by relatively frequent rains, nurtured by the hot summer sun, they stretch south into Montana.

Studded with Ponderosa pine and eastern Douglas-fir, this open woodland was used for millennia by the Ktunaxa, and in recent centuries was known as a perfect place for growing tobacco. It's unlikely anyone really knows how long native tobacco

(which is very different than commercial tobacco and was used very differently) has been grown in the region. What we do know is this: though it is now grown in many places around the globe, the origins of tobacco (the word comes from *tobago*, the name for a West Indian snuffing pipe) are tropical. Harvested as early as 6,000 years ago, the custom of using the dried leaves in a variety of ways for ceremonial purposes gradually spread both north and south from Central America.

The great differences in climate led to several species of tobacco being cultivated; none was like the modern, commercially grown plant that most people today associate with the word "tobacco". Among the species grown in North America are *Nicotiana ustica,* a bitter herb that was smoked through a pipe, *Nicotiana petunoicles*, a wild tobacco that was very harsh, and *Nicotiana quadrivalvis Pursh*, which was grown by the Ktunaxa. Some native cultures cut the harshness of the tobacco with kinnikinik, *Arctosta-phylos uva-ursi*, also known as bearberry, a broad-leaved, ground-hugging evergreen that grows virtually everywhere in Canada. Dried and then burned, kinnikinik burns with a lovely aroma that makes it clear why it was highly valued.

Photos: Ian Ward

Above right: A member of the lily family, beargrass grows to a height of 1.5 metres in open woodlands. Woven with cedar bark, an expert could make beargrass baskets that were watertight, without the use of sealants.

Horses still graze the meadows of the Tobacco Plains, just as their ancestors did at the end of the last glaciation.

By the time the first Europeans (who, by the way, were completely ignorant of tobacco of any kind) set foot in North America, tobacco was used almost everywhere for ceremonial purposes. Regarded as one of the handful of sacred plants (the others include sage, sweetgrass and cedar), it was generally smoked in sacred pipes, with the prayers of the people rising with the smoke to the Creator and to past generations of the people, but could be also used in offerings.

Tobacco was widely traded and many cultures used it to give thanks for success in hunting or gathering and to greet visitors from afar. Surprisingly, perhaps, tobacco was never smoked on a regular basis, as is the case in far too many aboriginal and non-aboriginal communities today.

Today, cattle roam the lush grasslands of this picturesque border region, while Rocky Mountain bighorn sheep—

Dennis Fast

Peter St. John

Elk twins follow their mother into the sheltering woodlands, while a bald eagle watches from the top of a ponderosa pine.

which are blue-listed (designated a vulnerable species) — graze the grassy upland meadows, elk and white-tailed deer are common along the forest edges and mule deer populate the grassy plains.

But the Rocky Mountain Trench—particularly its southern reaches—was once home to many other large mammals that are, alas, either extinct or extirpated from North America. According to the Royal B.C. Museum, these include mammoths, horses, burros, camels, musk-oxen and two species of bison. Though many sources contend that bison vanished from west of the Rockies after the last glaciation, archeologist

Wayne Choquette has recovered bison bones from the trench that date to as recently as 100 years ago. And Rexford Daubenmire, professor emeritus from Washington State University, reports that bison only became extirpated early in the nineteenth century. The open forest is also home to a large population of bald eagles, who nest in the pines and Douglas-firs.

Most of the trench was drier than present during the early post-glacial period, which began about 13,000 years ago, and grasslands dominated the landscape as pioneering plants colonized openings in the post-glacial ice. Pollen cores from Bluebird Lake southwest of Canal Flats, for example, indicate that sage-brush was an important part of the vegetation until about 6,600 years ago. Many of the current plant species migrated from the sagebrush steppe south of the glaciers in Washington, Idaho and Montana. Others immigrated through ice-free corridors in the foothills east of the Rocky Mountains.

In the past, fires caused by lightning were common in Ponderosa pine and interior Douglas-fir forests. Returning at intervals of between five and fifty years, they maintained the mosaic of grassland and open forest.

Getting There: The Tobacco Plains can be easily seen from almost anywhere along Hwy 93 from Elko south to the U.S. border. The Ktunaxa run a duty-free shop at the border and for those who would like to stay longer, the Edwards Lake Campground can be accessed via a road from the general store at Grasmere.

The Creston Valley

Tucked between the Purcell Mountains to the east and the Southern Selkirks to the west, the wide, flat Creston Valley slows the Kootenay River to a meandering crawl. Like the valley itself, the vast marshes, oxbow lakes and ancient channels are a legacy of the Kootenay River, which in turn is a descendant of Glacial Lake Kootenay.

For millennia, spring meant the rising of the Kootenay. Topping the river banks, the silt-laden water would spread across the valley floor, enriching the soil, renewing the wetlands and ultimately creating a floodplain that stretched from Bonner's Ferry, Idaho, through the Creston Valley to Kootenay Lake, a seventy-kilometre cornucopia of fertility and a haven for wildlife.

For millennia, the valley wetlands served as nurseries for the river's landlocked white sturgeon (see page 74) and kokanee salmon. The annual floods enriched the valley soils and rejuvenated its precious wetlands, nurturing thick stands of wild rice, and drawing migrating birds in numbers so great that they darkened the sky. Lined on both sides with lush upland meadows and dense hillside forests of cedar and pine, the valley was also home to great herds of elk and deer. Nearly 300 species of birds, from terns and great blue herons to tundra swans and Canada geese, either nested and bred in the marsh or used it as a key stopover point in their seasonal migrations.

Little wonder that the Ktunaxa have made this valley, along with the Goat and Moyie Valleys immediately to the east, their home for the past 11,000 years. In the mountains, archaeologists have found their ancient workshops. From Goatfell Quarry in the Moyie Range just east of Creston west to the elevated terraces along what were once the shores of Glacial Lake Columbia, early Ktunaxa developed a series of workshops that were used for thousands of years. Their preference for weapons and tools were quartzite and tourmalinite, stone that could be easily flaked by a practised hand.

Their large spear points, as well as knives and scrapers, indicate a lifestyle that centred on big game hunting. Archaeologist Wayne Choquette, the regional authority, believes that these early residents likely wintered in the valleys and spent the summers hunting and quarrying in the mountains. Though their relatives to the east hunted bison, the people of the Creston Valley were never short of prey.

On either side of the valley, five boreal ecosystems climb the slopes of the low, rounded mountains. Ranging from the warm, dry Interior Cedar-Hemlock subzone on the lower valley slopes to the mild,

The Kootenay River meanders through wild rice marshes in the Creston Valley.

dry Englemann Spruce Fir subzone at the upper elevations, the result was nirvana for ungulates. Huge herds of elk wintered in the valley, moose populated the marshes and wet meadows, white-tailed and

The bountiful marshes of the Creston Valley as portrayed by artist Henry James Warre in 1845.

mule deer grazed the grassy lower slopes, mountain sheep and goats were found on the uplands and mountain caribou, one of three subspecies of woodland caribou that live in B.C., thrived on the moss and lichen that carpeted the ground and hung from the branches of the old-growth trees. All this meant prime hunting grounds for cougars, wolves, wolverines and grizzly and black bears.

Little wonder the ancestors of the Yakan Nukiy—the "Meadow People" or Lower Ktunaxa—decided to stay so long ago. For millennia, the Creston Valley was nothing short of a paradise. As the centuries passed and the climate warmed, the Ktunaxa's quarry changed, but the people adapted perfectly. With their beautiful "sturgeon-nose" canoes, they slid through the marshes, harvesting wild rice and hunting ducks and geese, or went north to Kootenay Lake to fish for trout or the huge white sturgeon that once cruised the depths. In September, they headed for the many sparkling creeks and streams to intercept the freshwater salmon as they returned to spawn. Pitching their tipis at Kokanee Creek, which runs down from the Kokanee Glacier, or at Taghum Narrows, "the throat through which Kootenay Lake disgorges into the lower River", as D.M. Wilson describes it online in 'The Virtual Crowsnest Highway', they played host to an annual harvest feast.

And year-round, they bathed in the many hot springs scattered through the region and left a record in the hundreds of pictographs, or rock paintings, that line the shores of the lakes and decorate the cliffs above the valleys.

Change came with the first Europeans. Pushing the Ktunaxa from most of the land that was theirs for so long, draining and diking the wetlands and marshes, harvesting the old-growth forests, the settlers marked out their fields, planted their orchards and built their homes, their highways and ultimately their dams. All this had an enormous impact, not only on the Ktunaxa people, but on the thousands of species that for thousands of years had depended on this bountiful place and the river that eternally renewed it.

The Ktunaxa, who have never signed a treaty with any Canadian government, are currently negotiating a settlement. They can still be found in the Creston Valley and along the Kootenay River.

Yet even in among the early settlers, there were those who realized the value of the precious wetlands and the adjacent upland ecosystems. For almost a century, beginning in the early 1880s, one confrontation after another pitted developers, the

The striking western grebe once nested in ten places in B.C. Today, in addition to the Creston Valley and the south end of Kootenay Lake, breeding colonies are found only in Shuswap Lake and the north arm of Okanagan Lake.

provincial and federal governments and a host of business groups against the Ktunaxa, the Canadian Wildlife Service and Ducks Unlimited. Finally, in 1968, the Creston Valley Wildlife Management Act was passed. Now, the battle to restore the productive wetlands began in earnest.

Today, the 17,000-acre wildlife refuge holds the last remnants of many of British Columbia's species, including leopard frogs—this is the only place in B.C. that they're still found, though the Ktunaxa are involved in re-introducing the frogs to areas in which they once thrived; Forster's terns—this is their only breeding site in the province, and western grebes—the Creston Valley is one of only four places where they raise their young. This is also the site of the largest colony of great blue herons in the B.C. Interior, with more than 100 nests identified in the past decade.

Elk still gather in large herds, and deer and moose can still be found, but the region's mountain caribou have been reduced to two extremely vulnerable populations in the South Selkirks and South Purcells (see page 91). Though their range is technically outside the wildlife management area, the caribou's fragile hold on life in the region has recently spurred the provincial government to set guidelines for logging throughout their remaining range.

Restoration work also continues on other fronts, with slow, steady results. The resident population of

eagles and ospreys is growing; river otters and western painted turtles inhabit the reclaimed marshes, kokanee salmon spawn in Summit Creek and so many largemouth bass fill the warm shallow lakes at the south end of Kootenay Lake that fishermen are now a regular site where once developers envisioned fields of grain.

But challenges remain, including the desperate plight of the white sturgeon.

Getting There: Today, while many Ktunaxa live in Creston or other communities in the region, others live on reserve land along the river south of Creston. The band office, located in a building designed to resemble a tipi, offers guided trail rides into the mountains to the north, while the Ktunaxa Tipi Company manufactures tipis of many sizes. The community holds a powwow each year in May. In Creston, the museum, which has recently been renovated, has a collection of magnificent Ktunaxa points and tools.

To better understand the valley's ecology, visit the Creston Valley Wildlife Center, located just south of Hwy No. 3, approximately 11 km west of Creston. The Wildlife Centre has a boardwalk trail to a three-storey birding tower and an Interpretative Centre, which is open year round and features a hands-on display hall, naturalist-led nature programs, a nature film theatre, and a picnic area. Short, guided canoe tours are also available.

To visit the Creston Valley Wildlife Management Area, head west of Creston on Hwy 3 about 12 km and then turn north onto Dewdney Trail. Trails lead along Summit Creek and around Leach Lake. Or take Hwy 3A north of Creston to Lower Wynndel Road. Turn south for about a kilometre, west onto Duck Lake Road for 1.5 km then north on Channel Road to a parking area and trails around Duck Lake.

Dennis Fast

Sturgeon-nose Canoes

SHARPLY POINTED at both bow and stern, which were shaped like heads of the giant fish that once cruised the waterways throughout Ktunaxa territory, the flat-bottomed sturgeon-nose canoe was long unique to the Ktunaxa of the Lower Kootenay region. The unusual design, which extends the keel line to end "rams", is believed to help keep water out in rapid rivers and large lakes.

Sturgeon-nose canoes were traditionally made of birchbark, but also used the bark of spruce, fir, white pine or balsam, with cedar root or wild cherry bark used for the bindings. The canoes of the Ktunaxa represented the western-most use of birchbark for watercraft. For caulking, canoe builders used pitch from the ponderosa pine or Douglas-fir.

While this design was unique among North American canoes, craft of similar design have been used by the people of the Amur River region of Siberia for millennia. The sturgeon-nose design was used mainly by the Lower Ktunaxa of the Creston Valley and Kootenay Lake, but the design was adopted by the Shuswap from the north and the Okanagan of the southern Interior. The Upper Ktunaxa of the East Kootenays generally created dugout canoes or built rafts when they wanted to cross the rivers or negotiate the narrow lakes of the Columbia Valley region.

Sturgeon-nose canoes are still being built in the Creston Valley and a heritage example is on display in the Canadian Canoe Museum in Peterborough, Ontario.

These uniquely designed canoes were used on stretches of fast water, as well as the region's large lakes.

Peter St. John
Illustration, Dawn Huck

Mountain Caribou

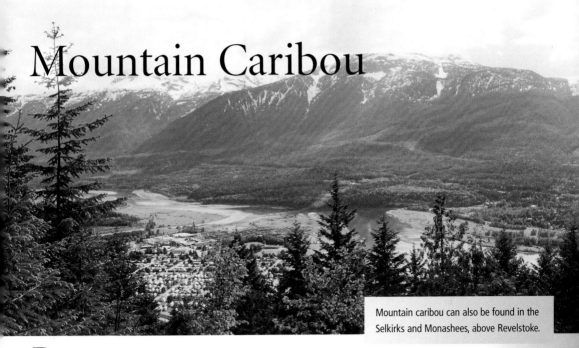

Mountain caribou can also be found in the Selkirks and Monashees, above Revelstoke.

DRIFTING like silent shadows through ancient forests, dancing on snow drifts as much as four metres deep, feeding on veils of tree lichen that hang from the branches far above the ground, mountain caribou were likely among the first animals to return to British Columbia following the last glaciation. And they have evolved to live in circumstances most large mammals find impossible. Yet today, in many places, they are rapidly disappearing.

Larger than mule deer, smaller than elk, mountain caribou are uniquely adapted for one of Canada's most difficult winter environments. Their hooves are their chief asset, for they spread like snowshoes, allowing them to walk on the surface of deep, hard-packed snow, rather than sinking into it as deer or an elk might. According to a paper produced by the B.C. Ministry of Environment, Lands and Parks, the hoof imprint of a mature caribou is about the size of a moose, yet the caribou weighs only half as much. Caribou also have exceptional coats, with hollow kinked hairs that trap a layer of warm air against their bodies.

A specialized type of woodland caribou (which are in turn one of five caribou species in North America), mountain caribou are so uniquely adapted to life in the old growth forests of British Columbia's interior wet belt—the world's only temperate inland rainforest (see map on page 93)—that ecologists have termed them a "flagship species". In short, mountain caribou might be found almost anywhere in their chosen environment, but not beyond it, for they have evolved a lifestyle quite different than that of their relatives, the boreal caribou of northeastern B.C. and the northern caribou of northern and west-central B.C. In both these regions the snow cover is much thinner.

The keys to their survival are the mountain caribou's twice-yearly vertical migrations. In early spring, the caribou head for the valleys, where they fatten up on spring's early greenery. They return to the heights when the snow melts on the alpine meadows, with pregnant cows seeking lofty, secluded locations that allow them to avoid predators during calving.

In late fall, when the winter's heavy snows begin, the caribou again migrate to the lower elevations where the closed canopy of old-growth forests shields the ground, allowing the animals to find forage, particularly mountain boxwood, a much favoured evergreen shrub.

Eventually the snowpack on the lower slopes

Treading lightly on a snowpack three or four metres deep, mountain caribou are elevated to a height where they can feed off the veils of lichen that hang from the lower branches of old-growth trees in the interior rainforest.

LF
06

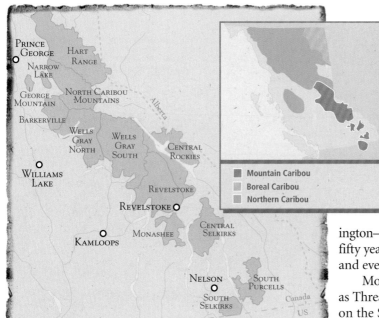

Mountain Caribou
Boreal Caribou
Northern Caribou

because ninety-eight per cent of the world's estimated 2,300 mountain caribou live in B.C.—the other two per cent are in northern Idaho and northeastern Washington—their decline in the past fifty years has caused national and even international concern.

Mountain caribou are listed as Threatened by the Committee on the Status of Endangered Wildlife in Canada (COSWIC), as are boreal caribou. Not only have they been extirpated in more than a third of their historic range, of the thirteen local populations identified in census information gathered during aerial surveys during late winter, several are in steep decline, and five herds, including the two most southerly, are at critically low levels with between five and thirty-five animals. A hard winter, avalanches, heavy predation, forest fires, disease, further harvesting of old growth forest: any or all of these could spell the end of one or more of these herds and further endanger the species as a whole.

Particularly at risk are mature bulls. Not only are they key to the survival of each herd, but because of the autumn rut, when they expend most of their time and energy on competing for mates, they often go into winter undernourished and weakened.

To try to stem the decline, the provincial government has initiated intensive studies of mountain caribou and has begun to put restrictions on both the harvesting of old-growth forests and the recreational use of key upper elevation range areas and adjacent lower areas that provide food for the herds in the early winter.

Canadians carry the caribou in our pockets and wallets, for these beautiful animals grace our 25-cent coins. It might be argued that inflation has greatly devalued the Canadian quarter since artist Emmanuel Hahn created the design in 1937, but surely the real thing is worth protecting.

deepens to the point that digging for food becomes impossible. About the same time, the deep snow on the upper slopes becomes firm enough to hold their weight and once again the caribou climb to the high mountain forests, where they spend the balance of the winter moving about on snow three or even four metres deep, and feeding on lichen that hangs from the branches of the oldest trees or blows off the branches during winter storms. Lichens alone provide the food they eat for the balance of the winter.

Not only does this high mountain environment provide the food they need, according to provincial government biologists it also creates a safety zone, because "the deep snow drives elk, deer and moose down to valley bottoms where there is less snow, and most predators follow them, leaving the caribou safely isolated."

It follows, therefore, that the harvesting of old-growth forests in the Columbia Mountains has had a disastrous effect on both the numbers and range of mountain caribou. And

Linda Fairfield

THE
THOMPSON
OKANAGAN

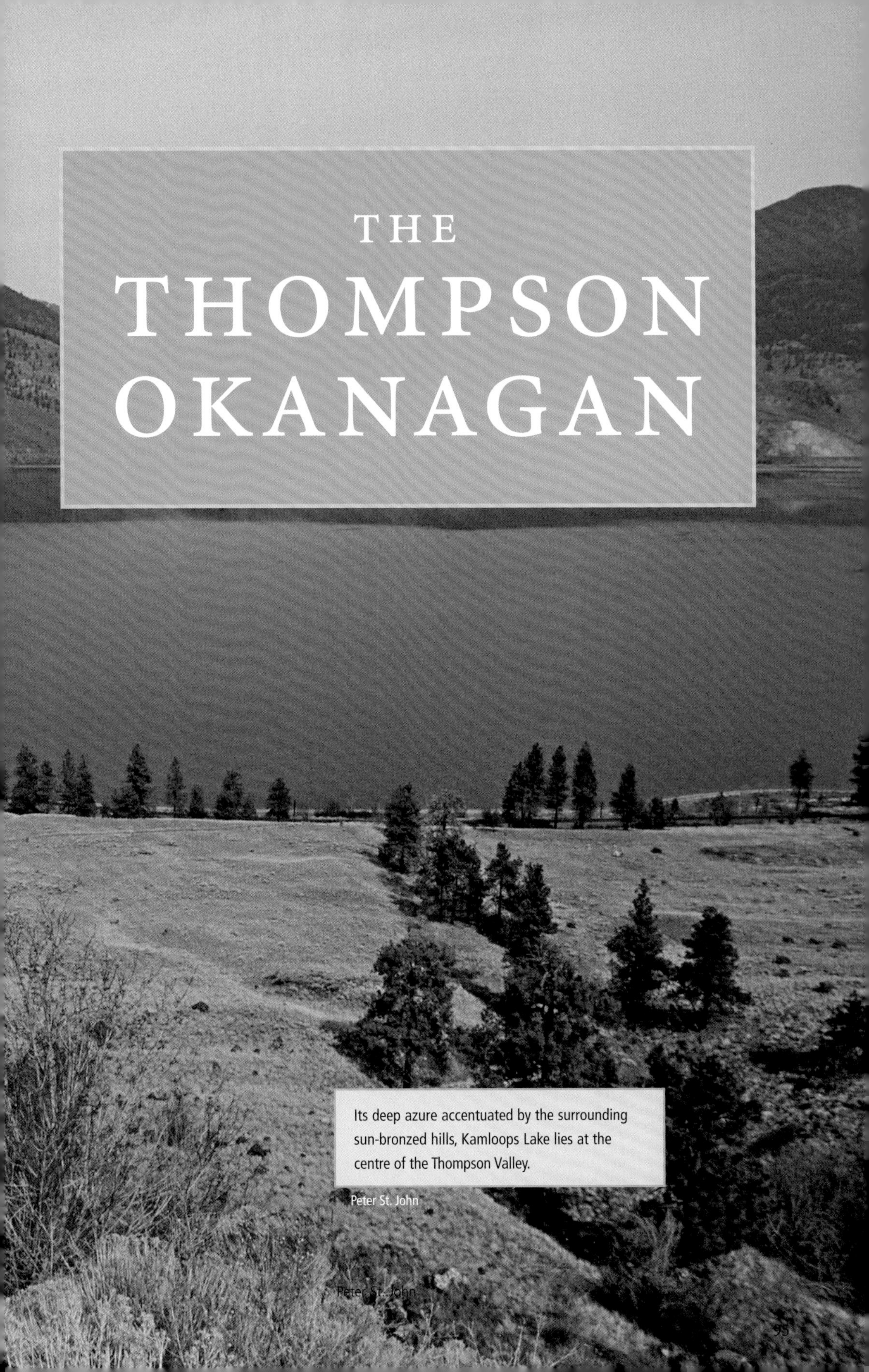

Its deep azure accentuated by the surrounding sun-bronzed hills, Kamloops Lake lies at the centre of the Thompson Valley.

Peter St. John

Peter St. John

At the southernmost reaches of the Okanagan Valley lies a small remnant of shrub-steppe, one of Canada's four most endangered ecosystems. The Nk'Mip Desert and Heritage Centre, shown here, is one of several places where concerted efforts are being made to save this precious natural environment.

The Thompson Okanagan

Travelling west from Revelstoke

to Kamloops and beyond is a revelation

of both the impact of elevation on precipitation

and the power of precipitation backed by climate change.

THOMPSON-
OKANAGAN
REGION

As THE TRANS-CANADA winds through the snow-capped Monashee Mountains beyond Three Valley Gap, it might seem to travellers encountering the wet, western slopes, with their dense cloak of fir and cedar forests, that the Pacific Ocean must be just ahead. Yet within hours, they emerge not onto the wild west coast, but rather into B.C.'s arid Interior, where sagebrush grasslands and ponderosa pine uplands are the norm.

The Thompson Okanagan region is filled with fascinating evidence of the enormous climatic swings the Earth has endured. The magnificent fossilized plants and animals of the McAbee Fossil Beds near Cache Creek, for example, preserve evidence of a bountiful period of global warming that began more than fifty million years ago (see page 112). And to the south, in the South Okanagan, two sites provide information about, as well as trails through, Canada's only desert.

At the other end of the temperature gauge, many places, including Deadman Valley at the west end of Kamloops Lake, along with the lake itself, attest to waves of global cooling and the region's glacial history (see page 104).

The Thompson Valley is also where Canada's oldest, securely dated human skeleton was found. Though the site is not marked, the incomplete skeleton of a young man, who was apparently caught in a mudslide, was found at Gore Creek, which flows into the South Thompson River near Kamloops. Excavated from sediments

Peter St. John

97

below a 7,700-year-old layer of Mazama ash, the remains were radiocarbon dated to 8,250 BP. Nothing in the way of artifacts was found with the skeleton, though elsewhere in the Thompson Okanagan, projectile points and other artifacts similar to those dated to between 8,000 and 10,000 BP in regions to the south and west have also been found. These have led archaeologists to believe that at least some of the area's earliest occupants were big game hunters who moved into the intermontane valleys in pursuit of deer, wapiti and caribou. Analysis of the Gore Creek skeleton showed that his diet was largely mammalian protein, not surprising in a culture that was largely focused on hunting, rather than fishing.

Since the slopes of the Thompson Valley consist largely of sediment brought here by the glaciers or deposited by the glacial lakes that marked their disappearance, landslides continue to be a frequent occurrence. A slide on the north side of the valley near Pritchard can still be clearly seen; the waterfall that plummets over the crest of the cliff is making its own changes to the topography. Downstream, where the valley walls are much more precipitous, at least nine large landslides have occurred since 1870 between Ashcroft and Drynoch, below Spences Bridge. One of the largest, in 1880, blocked the river for forty-four hours, forming a lake that extended fourteen kilometres upstream, according to Natural Resources Canada.

The region's human history is also evident in many places. Just beyond Pritchard, near the village of Monte Creek (where Highway 97 heads south into the Okanagan) a tiny provincial park protects an area of treed riverside grassland where traces of pithouses and underground ovens can be found. The kekulis or pithouses had entrances that faced the river. The park also contains a small portion of the Brigade Trail from the Okanagan, which dates to the era of the fur trade. The park has no day use or camping facilities.

Other sites, including Secwepemc Heritage Park in Kamloops (see page 106), and Nk'Mip Desert and Heritage Centre, in the South Okanagan (on page 128) provide an in-depth look at the people who have long made their homes in these arid landscapes.

In addition to Highway 97 at Monte Creek, the Okanagan Valley (see page 119) can be accessed farther east, via Highway 97A at Sicamous and from the south, where the Crowsnest Highway crosses Highway 97 just west of Osoyoos. Better known for its deep lakes, picturesque wineries and heritage orchards, the Okanagan is also geologically interesting, environmentally imperiled and home to one of Canada's most innovative first nations.

This chapter follows the main highways—in this case the Trans-Canada (No. 1) and Okanagan Valley (No. 97) Highways—from east to west and from north to south. If you are continuing west on the Trans-Canada, this section leads to the chapter on the Fraser Canyon on page 135. Those travelling east can continue with the section on the Kootenays on page 30.

Peter St. John

The vineyards at Nk'Mip just east of Osoyoos have recently produced some of Canada's best wines. And the wine-tasting facility is second to none.

The Secwepemc

IT WAS VERY LIKELY salmon that allowed the distant ancestors of the Secwepemc, or Shuswap people, to create first villages on the Interior Plateau. Situated in what is now Shuswap Lake Park, the village was on the north shore of the lake just east of the mouth of the Adams River, perhaps the most remarkable salmon spawning site in B.C. Closer to the village, Scotch Creek was then and is today also a salmon spawning site. It's not surprising, therefore, that using techniques that ultimately included weirs, dipnets and leister spears, the Shuswap Lake people became masters at harvesting them. Dried in the region's warm winds, salmon enabled the Secwepemc to abandon their mobile lifestyle and settle in villages.

Anthropologists also believe that a period of global cooling, beginning about 4,500 years ago at the end of the Hypsithermal, may have assisted in making the idea of warm, permanent homes more appealing.

Whether it was a dependable food source, the cooler, wetter winters, or both, a permanent village of large kekulis, or pithouses, grew up near Scotch Creek—as they did in dozens of other places—about 4,000 years ago. Used mainly during the fall and winter months, the structures averaged more than ten metres across with rectangular floors dug into the earth and steep walls. Not surprisingly, they also had massive roof structures.

Though some villagers—usually elders and mothers with young children —stayed in the village year round, many others spent the spring and summer in the surrounding forests and hills. There they hunted elk, deer, and mountain sheep, as well as ducks and geese, and gathered berries and plants for use as food, to make medicines, and for dozens of items from mats for roofing and flooring to baskets and clothing. They also fished for trout and collected freshwater mussels.

In such a bountiful place, it's easy to understand why the village was inhabited for more than 2,000 years. Every decade or two, the houses were rebuilt, and over time, their inhabitants adopted new technologies, used new materials and embraced new artistic styles. About 1,500 years ago, hunters moved from using the atlatl—a spear thrower, tipped with a medium-sized stone point on a detachable shaft—to the lighter bow and arrow, which allowed accuracy and lethality at a greater distance. As the centuries passed, the quality of the workmanship of the projectile points, knives and scrapers increased. And the quality of the stone itself improved, indicating wider trading networks and better information on sources of good quality stone.

The passage of time and the impact of other cultures was also reflected in increasingly sophisticated art; here and elsewhere in the Interior Plateau, beautifully carved artifacts of stone, bone and antler appeared, along with harpoon points barbed on one or both sides, a clear indication that new ideas were emerging, some from as far away as the coast.

SHUSWAP VILLAGES
△ 4000–2400 BP
▲ 2400–1200 BP
▲ 1200–PRESENT

The Marquis of Lorne, Canada's governor general, sketched Little Shuswap Lake, below, during a trip across Canada in 1881.

Archaeologists have now excavated sites along the North and South Thompson, middle Fraser, Salmon and Shuswap Rivers, as well as on the shores of the Shuswap, upper Okanagan, Nicola, and Arrow Lakes and have divided the long history of the Secwepemc people into three periods or "horizons": the Shuswap horizon (4,000 to 2,400 BP); the Plateau horizon, (2,400 to 1,200 BP) and the Kamloops horizon, between 1,200 BP and the arrival of European settlers less than 200 years ago.

Everywhere there was the same dependence on salmon, the same otherwise varied diet, and the same changes in the design of the kekulis, ultimately resulting in houses that were widely varied in size, with lighter roofs and side entrances. Near many of the most recent houses were small circular depressions; some were used for storage, while others were earth ovens, used for roasting food.

Not surprisingly, given that in many villages the houses were rebuilt again and again, artifacts from the various periods became jumbled over time. But archaeologists are aided in dating sites by the same progressions in the style of artifacts and kekulis in many places.

Excavating the Shuswap Lake site in 1972, however, archaeologist John Sendey found something quite unusual beneath the floor of one of the earliest houses: the bodies of not one, but eight people. Though they were interned without grave goods, ochre had been liberally sprinkled over the bodies. Such a mass grave arouses all kinds of questions in a place where infectious diseases were all but unknown. Were these the victims of an attack? Did they die of food poisoning, perhaps? Or starvation, when the salmon failed to come?

Eventually, perhaps because it was easier to build new homes than continue to repair the old ones, the Shuswap Lake site was abandoned. But the people didn't go far; villages dating to between 1,200 years and the arrival of Europeans in the area have been located west of the earlier site, as well as on the south shore of Shuswap Lake and the west shore of Adams Lake near Bush Creek in what is now Adams Lake Provincial Park. And today the Secwepemc people still live along Scotch Creek, little more than a stone's throw from the land on which their forebears lived for so long.

The Adams River

TUMBLING SILVER and white from Adams Lake through a canyon of its own making, rushing sea green between the riverside forests, swirling through the spawning channels and over the gravel-bottomed pools that line its mouth, for generations, the Adams River has awaited the coming of vast numbers of spawning sockeye salmon.

Today, though tour groups continue to promote the spawning spectacle of the Late Shuswap run, which takes place largely during October in dominant years—2010, 2014 and every fourth year following—there is growing fear here and elsewhere along the Fraser River system, that overfishing, habitat degradation and salmon farming are combining to destroy a species upon which British Columbians have depended for millennia.

Salmon runs are not all equal; far from it. Barring the lengthy consequences of natural disasters, for the past six or seven thousand years—since salmon runs were re-established in the province's rivers following the last glaciation—every fourth year has produced a dominant run, during which numbers spike with more than four times the number of spawning fish than can be seen in the next largest, or subdominant, year. During the other two years, only small numbers of sockeye climb the Fraser and the North and South Thompson to the hundreds of tributaries, some more than 500 kilometres from the ocean, where the salmon spawn. Though 485 kilometres distant from the mouth of the Fraser—a seventeen-day battle for healthy sockeye salmon—the Adams River has produced the single largest spawning population in the Fraser River watershed.

Through most of the twentieth century and the early years of the twenty-first, however, conditions were anything but optimum for the Fraser River sockeye. In large part, this was because of the disastrous 1914 Hell's Gate landslide in the Fraser Canyon (see page 150). Triggered by the construction of the railway through the canyon, the slide blocked the river and all but destroyed the fishery. Through the balance of the century, its comeback was slowed by habitat destruction, commercial overfishing, pollution in the lower Fraser River and occasional extremes in water temperatures or water levels.

Little wonder that the revival of the dominant sockeye run in 2006 caused huge excitement among fisheries specialists, university scientists, fishermen of all description and the thousands who simply want to witness one of the world's great natural events, something that rivals the migration of Africa's wildebeest.

However, since then, the news about B.C. salmon has been virtually all bad. Fraser River runs in 2008 and 2009 were so small (the latter was just one-tenth the size that had been predicted) that commercial fishing was all but completely curtailed and the federal government established a Commission of Inquiry into the dramatic decline.

The small run in 2008 was not completely unexpected. A report released in 2005 by the House of Commons Fisheries Committee predicted that the

In addition to the sockeye in the huge Late Shuswap run, chinook and coho salmon also once spawned in this remarkably bountiful stream.

Peter St. John

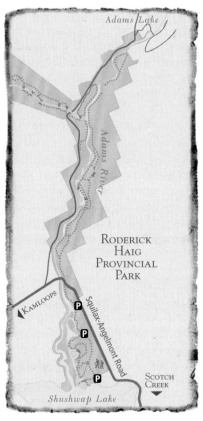

sockeye fishery on the Fraser River could be wiped out in 2008. A combination of a poor yearly run in 2004, high water temperatures in both the Fraser and some of its tributaries and an unauthorized and unreported harvest combined to greatly diminish the number of salmon available to spawn. The only option, according to the committee, may be to cancel the quadrennial sockeye fishery until the run recovers, which might take until 2020.

The failure of the run in the fall of 2009, however, was completely unexpected, and not only British Columbians, but Canadians across the country could talk of little else as the autumn progressed and the salmon failed to appear. Declaring the sockeye "commercially extinct", many compared the disaster to the disappearance of Atlantic cod stocks more than a decade before. However, it's unlikely that falling numbers of sockeye mean more to any group than they do to the Secwepemc, who have lived along the shores of Shuswap Lake and Adams Lake, which feeds the Adams River, for at least 4,000 years. Salmon have been the mainstay of Secwepemc culture since long before the pyramids of Egypt were built. And even today, according to fisheries experts Evelyn Pinkerton and Martin Weinstein, "about 57% of all Fraser River sockeye salmon, as well as 25 to 34% of Fraser River Chinook and Coho salmon respectively" spawn in Secwepemc territory.

Though some are still pinning their hopes on the dominant sockeye run, it is clear that like other salmon species elsewhere—from the huge "June hogs" that once filled the late spring waters of the Columbia River to the chum salmon that have largely disappeared from the streams along B.C.'s north coast—salmon, for so long a defining factor in the province's lives, are in a dire state of emergency.

Getting there: Roderick Haig-Brown Provincial Park is located on both sides of the Adams River, between Adams Lake and Shuswap Lake. From the Trans-Canada west of Sorrento, take the Squilax-Anglemont Road north over the bridge and follow the road across the Adams River to the provincial park and the hiking trail along the river. To see the spawning salmon, turn east immediately after the bridge and follow the road south. The Lower Trail System allows access to the spawning channels; parking is provided in several lots. Spawning generally occurs between early September and mid-October.

The Thompson River

TODAY'S THOMPSON River flows south and west; its north branch tumbles from headwaters in the glaciers of the Cariboo Mountains west of Jasper and curves south of Wells Gray Provincial Park, while the South Thompson begins in the upper reaches of the Monashees and widens through the Adams and Shuswap Lakes, as well as Little Shuswap Lake before slowing to join its northern sibling at Kamloops. Continuing west through Kamloops Lake, the descendant of glacial waters that filled the valley for thousands of years, the main Thompson River flows through the arid grasslands of the southern Interior. Turning south at McAbee, known for its magnificent Eocene

follows the South Thompson from Sorrento, at the south end of Shuswap Lake, to Kamloops and then traces the main river to its meeting with the Fraser.

But geologists believe the Thompson was not always a tributary of the Fraser. While the lower Thompson Valley from Kamloops to Spences Bridge is at least fifty million years old; for much of that time, geologists believe, the river may have flowed northeast rather than southwest. From headwaters that were likely somewhere between Hell's Gate and Boston Bar, water from many valleys, including the Stein and Nahatlatch Rivers, flowed into a northeastward trending drainage that may have been the headwaters of the ancestral Peace River.

It is also possible that the North Thompson and perhaps even the headwaters of the South Thompson may have flowed northwest (instead of southwest) through ancient valleys across the Shuswap Highlands and Cariboo Plateau to join a large river near Quesnel.

Peter St. John

Wave-cut benches from the glacial lakes that once filled the valley can be clearly seen along the hillside in this view of Kamloops Lake.

fossils (see page 114), it rolls through the deep quiet canyons and thunders over the powerful rapids of its lower reaches to its confluence with the Fraser. The Trans-Canada Highway

This northwest-trending watershed may have lasted for almost fifty million years, perhaps until well into our current ice age, which began about two million years ago.

All of this, as indicated on page 159 means British

Peter St. John

Columbia's famous Fraser River is a "pirate". Aided by the uplift of the Interior Plateau as well as glacial and fluvial erosion, it has stolen water from an enormous region of B.C. over the recent past, geologically speaking.

Today, the Thompson River is both the longest tributary of the Fraser, draining a watershed of more than 55,000 square kilometres in central B.C., and the largest contributor to the waters of the lower Fraser. Fully one quarter of the water of the lower Fraser comes from the Thompson River.

In fact, it may be the Thompson River (or perhaps more properly its two glacial lakes—Glacial Lake Thompson and Glacial Lake Deadman) that removed the last of the obstacles preventing the Fraser River from staking its claim to its enormous watershed. Of the two, Glacial Lake Deadman was the real perpetrator, but more on that later.

Though geologists are still wrestling with their glacial timeline as a result of the presence of salmon fossils that have been radiocarbon dated to between 15,000 and 18,000 BP (or between 17,940 and 21,390 in calendar years) (see page 108) along the south shore of Kamloops Lake, there seems to be agreement that the main valley of the Thompson River contained at least four stages of two main glacial lakes following (and perhaps even prior to) the height of the last glaciation. All were long, ribbon-shaped lakes that collected glacial meltwater and sediments as the ice decayed on the Interior Plateau. And all, except the last one, drained to the east and south. According to geologists Timothy Johnsen and Tracy Brennand, whose recent paper on the late-glacial lakes in the Thompson Basin was published on the National Research Council website, glacial lakes are given new names "when (1) the lakes are in separate locations or (2) the lake outlet position changed significantly." The latter was the case with the lakes of the Thompson Basin. Glacial Lake Thompson, the earlier and deeper of the two, drained through an outlet near Chase, just west of Little Shuswap Lake, while until its last, catastrophic drainage, or *jökulhlaup*, Glacial Lake Deadman had its outlet just east of Kamloops.

Wave cut benches—distinct horizontal terraces along the hillsides throughout the valley—show that at its highest stage, Glacial Lake Thompson was at least 140 metres deep, while at its lowest stage, Glacial Lake Deadman was about fifty metres deep.

Unlike the huge Laurentian ice sheet that actively retreated from Canada's heartland at the end of the last glaciation, the valley glaciers of the Cordilleran ice sheet melted in place, wasting away in the valleys, while persisting at higher elevations. Thus stranded in upper passes, these dead ice masses formed ice dams that eventually gave way, allowing each stage of the glacial lakes to drain. Filling again with meltwater from the valley ice and from adjoining valleys, each of these stages may have existed for as little as eighty years or as long as 1,130 years, based on radiocarbon studies.

An arm of the glacial lakes filled the Deadman Valley, opposite. Drained repeatedly, the rushing water created the valley's upper canyon; the Thompson River canyon, below, was very likely created by the *jökulhlaup*—Icelandic for "glacier burst"—that drained Glacial Lake Deadman.

As the ice decayed, the lake levels lowered and lengthened to the west until, sometime between 9,200 and 10,800 BP in radiocarbon years, an ice dam at the western end of Glacial Lake Deadman suddenly failed, discharging a huge wall of water—up to twenty cubic kilometres. Pouring southwest, this devastating flood may have triggered the failure of other glacial lakes downstream and perhaps even drilled through a height of land that separated the headwaters of the ancient, northflowing Peace River from the headwaters of the much smaller Fraser.

Roaring down the Fraser Canyon, the floodwaters reached the Strait of Georgia, a distance of about 250 kilometres, and swelled—tsunami-like—across it. It may even have been this ice dam failure that was responsible for the deposits that scientists have found in Saanich Inlet on Vancouver Island (see page 235).

Both Glacial Lake Thompson and Glacial Lake Deadman deposited enormous quantities of sediments, washed from the Interior Plateau, into the Thompson Valley. Johnsen and Brennand estimate that glacial action had created a valley up to 1500 metres deep; sediments from the glacial lakes filled it with silt and sand up to 800 metres thick. In the 12,000 years since, water and wind have incised these sediments, leaving obvious terraces or wave-cut benches, along the valley sides between Kamloops Lake and Spences Bridge, as well as hoodoos in a number of places.

Spectacular rock hoodoos, eroded from deposits of volcanic ash, as well as yellow cliffs and lovely waterfalls, can be found in the Deadman Valley north of Kamloops Lake.

Getting there: Wave cut beaches can be seen in many places along the Thompson River Valley between Kamloops Lake and Spences Bridge, including the lookout near Savona. These long ridges can also be seen along the north side of the valley where the Deadman River comes into the Thompson Valley 75 km northwest of Kamloops. To see the Deadman Valley hoodoos, turn north from the Trans-Canada Highway onto the Deadman-Vidette Lake Road, about 2 km west of the bridge over the river at the end of Kamloops Lake. The road runs through the Skeetchestn Reserve; the village of Skeetchestn offers Secwepemc cultural weekends, including camping in kekulis or traditional pithouses. The land around the hoodoos is a protected area; camping is not permitted there, but recreational sites to the north and northwest allow tenting.

Artist unknown / National Archives of Canada / C-104441

105

Kamloops

Photos: Peter St. John

Kamloops—or more properly *cumloups*, "the meeting of the waters", in the Interior Salish language of the Secwepemc people who have lived here for thousands of years—sits at the junction of the North and South branches of the Thompson River. Here, just above the south branch of the river and east of the confluence, near what is now Secwepemc Heritage Park (see below), the Secwepemc people have lived for thousands of years. Though today this land is dry, for millennia much of it was wetlands and around this marsh a large community grew up. As recently as the middle of the twentieth century, more than 200 depressions could be seen. These were the remains of pithouses, many of which had been occupied for thousands of years. And above the site, on a large glacial terrace, artifacts dating back to a little-understood period between 11,000 and 6,000 years ago have been found.

Every summer for more than a decade, George Nicholas, a professor of archaeology at Simon Fraser University, directed a field school that was jointly sponsored by the Secwepemc Cultural Education Society and SFU. The summer school projects not only introduced hundreds of aboriginal and non-

aboriginal students to a hands-on post-secondary archaeology program, but also uncovered indisputable proof that the lower slopes of these low mountains—which today are known as Mount Paul and Mount Peter—have been regularly occupied for more than 6,000 years and very likely used as encampments or fishing sites long before that.

Among the hundreds of projectile points and tools found during the excavations of several sites on the terraces, was a leaf-shaped spear point more than six centimetres long, similar to those used across much of North America between 11,500 and 10,000 BP. Crafted of chalcedony not found in the region, it was an ephemeral sign from the distant past. Another chalcedony projectile point, also found on the terrace, had been notched on both the base and the sides; it is unlike anything yet found anywhere else on the Interior Plateau. Also excavated were tiny microblades, slivers of sharpened stone that would have been hafted into a handle, rather like a sharp scalpel. At first glance, this might seem an unusual place to have attracted such far-flung and long-term attention.

Secwepemc Heritage Park includes reconstructions (above) of the kekulis or pithouses that long occupied the site, as well as partial houses that allow the structure to be seen. The museum's displays include this unusual curved scraper.

Dennis Fast

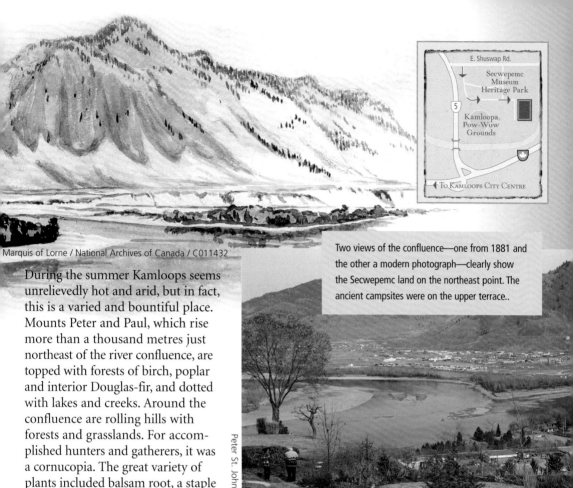

Marquis of Lorne / National Archives of Canada / C011432

Peter St. John

Two views of the confluence—one from 1881 and the other a modern photograph—clearly show the Secwepemc land on the northeast point. The ancient campsites were on the upper terrace..

E. Shuswap Rd.

Secwepemc
Museum
Heritage Park

5

Kamloopa
Pow-Wow
Grounds

← TO KAMLOOPS CITY CENTRE

During the summer Kamloops seems unrelievedly hot and arid, but in fact, this is a varied and bountiful place. Mounts Peter and Paul, which rise more than a thousand metres just northeast of the river confluence, are topped with forests of birch, poplar and interior Douglas-fir, and dotted with lakes and creeks. Around the confluence are rolling hills with forests and grasslands. For accomplished hunters and gatherers, it was a cornucopia. The great variety of plants included balsam root, a staple food, and saskatoons—"real" or "ordinary" berries in the Secwepemc language—which were eaten fresh or dried like currants.

And all this is without mentioning the salmon; more than half the total Fraser River spawning run of sockeye and between a quarter and a third of all the chinook and coho go right past the confluence on their way to the spawning streams of the Shuswap Lakes. And not surprisingly, given the warm winds of the Thompson Valley, some of the best dried salmon came from the Interior.

It was undoubtedly the people's ability to harvest large quantities of salmon that changed the seasonal campsites on the glacial terraces into semi-permanent settlements almost 4,000 years ago. One of the most important was the village just up the South Thompson River from the confluence. Excavating there, archaeologists have found evidence of all the cultural traditions of the Interior, from the Shuswap horizon beginning about 3,800 BP to the arrival of Europeans in the valley 200 years ago.

Today, the people who have lived here for so long are celebrating their history with the Secwepemc Museum and Heritage Park, which aims to preserve their language, history and culture.

The five-hectare park includes displays on virtually every aspect of life: more than a kilometre of trails wind through the archaeological remains of a 2,000-year-old village; four reconstructed winter pithouses; a summer mat-lodge village; a hunting lean-to, fish trap, drying rack and smoke house, as well as ethno-botanical gardens that represent the ecosystem of the region. Like a time machine, the park easily transports visitors back in time.

Getting there: The Secwepemc Museum and Heritage Centre is just east of the Kamloopa Pow-Wow Grounds on the northeast shore of the confluence. From the Yellow-head Highway, follow the map, above.

The Fossil Salmon of Lake Kamloops

SALMON have been crucially important in the development of British Columbia's human history. But salmon are also playing a role in determining the province's geological history and glacial history. Fossils of the world's earliest known salmon species, the rather lyrically named *Eosalmo driftwoodensis*, found along Driftwood Creek near the northern town of Smithers (as well as near Princeton and in Washington State), have also been of assistance in piecing together Canada's climate history (see page 16). And much younger salmon fossils—a mere 15,000 to 18,000 radiocarbon years, or 17,940 to 21,390 calendar years old—may yet assist in sorting out the province's glacial history.

Eosalmo dates from a dramatic period of global warming—the Eocene Climatic Optimum—about fifty million years ago. This was a period of exploding diversification of flowering plants, and the development of the earliest ancestors of many trees found in British Columbia today, including birch, maple,

willow, pine and fir. In fact, so warm was the Earth that in Canada's far north, on Ellesmere Island in the Canadian Arctic (which was even then above the Arctic Circle), whole forests have been found, still rooted in the ancient soil in which they grew. These include fossils of trees more than a metre in diameter.

The Earth's remarkable Eocene warming (which makes today's rise in global temperatures seem trivial by comparison) is discussed further on page 112, but much more recent fossils of salmon, found along the south shore of Kamloops Lake pose some fascinating questions about our last period of deep global cooling— the Wisconsin glaciation.

These fossil salmon were initially

Peter St. John

Viewed from a picnic area near Savona, Kamloops Lake stretches serenely east, but its quiet surface hides many secrets.

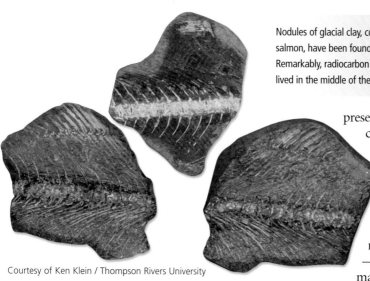

Nodules of glacial clay, containing the remains of kokanee salmon, have been found along the shore of Kamloops Lake. Remarkably, radiocarbon dating has shown that the salmon lived in the middle of the last glaciation.

Courtesy of Ken Klein / Thompson Rivers University

found in more than thirty years ago by the late geologist and paleontologist Richard Hughes, in deposits of fine-grained glacial clay—rather like potter's clay—along the south shore of the lake. Preserved in lumps shaped like the salmon remains they contain, the bones are recognizably Pacific salmon. Some even have the prominent hooked jaws that develop during spawning. Yet Carbon-13 measurements show that these fish never consumed food from marine sources. And radiocarbon dating have put the fossils at between 15,480 and 18,0110 years old; in short, these are kokanee salmon, but the dates have them swimming in Kamloops Lake when much of B.C. was covered by ice.

Hughes initially believed the fossils were "Late Pleistocene"—fish that lived at the end of the last glaciation—for he found them, along with many cannonball-shaped lumps of pure clay, in glacial sediments. For decades, that seemed hard to believe, for conventional wisdom had it that significant numbers of salmon didn't repopulate British Columbia's rivers until perhaps 6,000 or 7,000 years ago.

Then, in the early 1990s, anthropologist Catherine Carlson and paleontologist Ken Klein began to study the fossil site. As might be expected, they discovered that the best preserved fossils were those with concentrated bone: skulls, backbones (some with ribs attached), as well as basal portions of the fins and tail. Only one specimen appeared to be almost complete; the rest were pieces of salmon —likely scavenged by animals so long ago, as is the case with spawning salmon today.

Inside the clay nodules or concretions, the skeletons were remarkably intact, with bone structure that was easy to see. Moreover, the skeletons had not completely mineralized and therefore contained organic bone that could be dated. The first specimen collected, from the surface of the shore, returned a radiocarbon date of 15,480 years. Since everyone accepted as gospel that the Thompson Valley was full of ice 15,000 years ago, the scientists felt that the sample must have been contaminated.

Digging into the glacial sediments, they found buried—and therefore presumably uncontaminated— fossils and sent them to be dated. One can imagine the shock when these came back with an even greater age—more than 18,000 radiocarbon years (or more than 21,000 calendar years) old.

The implications are startling. Was the Thompson Okanagan region—like southern Vancouver Island— not glaciated until very late? Or was it filled even 18,000 or 20,000 years ago with glacial meltwater? Carlson believes the latter to be the case. With the drainages in the Thompson River flowing east and south at the time, and connected to the unglaciated Columbia River, salmon would have been able to migrate into the Thompson meltwater.

It's a stunning thought, for if these waterways were suitable for freshwater salmon, wouldn't they have attracted other animals as well? What, after all, had scavenged the fossilized salmon? And if animals were living around or along these lakes and rivers, might humans have also been there?

Clearly, as geologists studying North America's other great glacial lakes have found, they give up their secrets only reluctantly.

Global Warming?
Or Another Glaciation

WITH SUCH OBVIOUS INDICATORS as Earth's melting global ice caps continually in the news, global warming has become a hot topic of conversation. Several international projects recently completed by NASA, as well as American and Canadian universities, suggest the past may offer some clues to predicting what lies ahead. Yet at least one study has come to conclusions that are "polar opposites" of the others.

Dennis Fast

About fifty-five million years ago, at the beginning of the Eocene epoch (or, for dinosaur buffs, about ten million years after the end of the Cretaceous), the Earth suddenly rapidly warmed. Over a period that was likely considerably less than 5,000 years, the global temperature climbed between four and seven degrees Celsius. By comparison, global temperatures have increased just over half a degree since the Industrial Revolution. And two of the three warmest periods since the last glaciation were the Hypsithermal, which reached a peak between 8,000 and 5,000 years ago and the Medieval Warm Period,

which allowed Norse farmers to settle Greenland between 1,200 and 700 years ago (or 1300 and 800 AD). In both cases, global temperatures were warmer than they are today, but considerably cooler than they were during the early Eocene.

Global warming is connected to an increase in greenhouse gases, principally water vapour, carbon dioxide (from auto exhaust, wood smoke, animal exhalations and decomposing vegetation), methane (mainly from the stomach gas of ruminants, the decay of plants and animals and from natural gas) and nitrous oxide (mainly from decomposing manure, chemical fertilizers and even from catalytic converters). Though carbon dioxide (CO_2) is deemed to be responsible for about fifty-five per cent of the climate change over the past century, it is methane and nitrous oxide that are causing more concern, for they have, respectively, more than twenty times and nearly 300 times the potential of CO_2 to cause climate change.

Methane, scientists believe, was largely responsible for the abrupt global warming that took place fifty-five million years ago. A dramatic change in the Atlantic Ocean's circulation systems operating at the time caused a mass extinction of deep-sea dwelling marine life, which in turn released stores of methane gas that accelerated the warming process, further changing the oceans' circulation.

In a study published recently in *Nature*, scientists from the Scripps

© W.S. Thomson, Edinburgh, Scotland

The Northern Hemisphere has known remarkable climatic extremes, from enormous glaciers, opposite, to subtropical forests. Today, thanks to the Gulf Stream, palms thrive in Scotland's Inverewe Gardens, above. But no one is certain what lies ahead.

Institution of Oceanography focused on the Eocene global warming and used it as an example of the changes a massive release of greenhouse gases could trigger today.

Coincidentally, studies by NASA and Canadian scientists have shown that today's counterclockwise current in the North Atlantic—commonly called the Gulf Stream but technically known as the sub-polar gyre—has significantly weakened since the late 1990s compared to the 1970s and 1980s. This is the current that warms Britain and Northern Europe (allowing, for example, palm trees to grow on Scotland's west coast). Clearly, a slowing or even an end to it would dramatically change the climate in the North Atlantic. But studies undertaken by international specialists, including British Columbians, believe

the result would more likely be global cooling than global warming.

The gyre, which can take up to twenty years to complete its route, carries warm, buoyant water northward from the Caribbean along Europe's west coast, before turning westward past Ireland. Losing heat as it goes north, the water warms Northern Europe before becoming cold and dense. Plunging to the salty depths, it begins its slow journey south toward the equator. To modern eyes, the Gulf Stream seems a climatic constant, but in fact winter storms, as well as the buoyant fresh water pouring off melting glaciers all have the ability to impact it and the potential to slow it or even bring it to a halt.

Lead NASA researcher Sirpa Hakkiner believes "[The slowing of the current] is a signal of large climate variability in the high latitudes." And others have painted the following scenario: a continuation of our global warming; the accelerated melting of the polar ice caps; a resulting flood of fresh water; a slowing or even an end to the sub-polar gyre and the beginning of another glaciation (see The Icing on British Columbia on page 16).

Climatologists are using computer models and biogeochemical interpretations, as well as longterm studies in the North Atlantic to try to predict or better understand the triggers for global climate change. But is it possible the answer is buried deep in the Okanagan Highlands?

The Okanagan Highlands

This lovely view of the Spillimacheen Valley was rendered by John Douglas Sutherland Campbell, the Marquis of Lorne and Canada's governor general from 1878 to 1883, on a trip across Canada in 1881.

THOUGH THEY'RE NAMED for the Okanagan region, in fact, these ancient volcanic highlands stretch from northeastern Washington State, along the west side of Okanagan Lake nearly 1000 kilometres north to Driftwood Canyon, near Smithers. Their deposits of coal and lava, which outcrop in a number of places along this broad swath, recall one of the warmest periods in Earth's long history. The Eocene, between 55.5 million and 33.7 million years ago, was well named, from the Greek words *eos*, meaning "dawn" and *ceno*, meaning "new", for this was indeed the dawning of a new world, the world we know today.

Imagine a period when the Canadian Arctic boasted not glaciers (nor polar bears), but trees with trunks more than a metre across. Imagine B.C. forests that more properly belong in the subtropics, winters with hardly a day of frost, and an explosion of life that set the scene for the rest of time.

Geologically, this was a time of thrusting and stretching, as the Earth's tectonic plates began to change direction. The headlong movement of the Pacific Plate into and under the North American plate slowed, and other smaller plates began to slide against the continent, releasing some of the pressure and creating crustal faulting, basins and fractures. This gave birth to volcanoes; from B.C.'s interior plateau south into Washington State, their eruptions spewed lava over huge areas, building the Interior Plateau and damming rivers.

In places, such as the Princeton Basin, large depressions were quickly filled with sediments, perfect for the formation of coal swamps. Elsewhere, like the Okanagan Valley, half-grabens, deep faults that served as doorways into the Earth's mantle, filled with water. Everywhere, life was changing in a world no longer dominated by dinosaurs and in many places, those changes were preserved by the volcanic sediments and layers of lava.

Today, in at least seven sites in British Columbia, as well as places as far flung as Florissant Fossil Beds National Monument in Colorado and Ellesmere Island in Canada's far north, the Eocene is attracting a great deal of attention. In part, that may be because the modern world is suddenly very interested in global warming, but the Eocene is fascinating in its own right, for the fossils of fish, plants

EOCENE HIGHLAND LOCALITIES

Driftwood Creek, Smithers

Quesnel

Fraser

Horsefly

Cache Creek & Kamloops Area (McAbee Fossil beds)

Nicola & Quilchena Area

Okanagan

Princeton & Coalmont Area (incl. Similkameen & Tulameen Rivers)

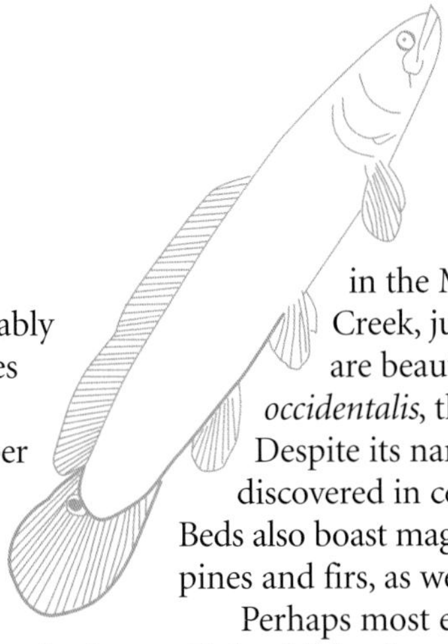

and animals that are being unearthed are very recognizably the ancestors of many species that exist today.

Among them is a member of the bowfin family. Found near Princeton, these ancient fish, which originated almost 100 million years ago, can still be found today in several of the smaller Great Lakes, as well as in slow-moving rivers like the Mississippi. Unlike most other modern fish, the bowfin—which is also known as the dogfish or mudfish—can extract oxygen from the air and is able to survive for a time in the mud below dried up lakes and flooded rivers. *In Fishes of the Central United States,* biologist and artist Joseph Tomelleri and his co-author, biologist Mark Eberle, tell of Mississippi Valley farmers "who occasionally turned up live bowfins with the plow after floodwaters had receded from farmland."

Farther north, ancestral salmon, *Eosalmo driftwoodensis,* were found along Driftwood Creek, a magnificent Eocene site near Smithers.

Other fascinating fossils, found in the McAbee Fossil Beds near Cache Creek, just north of the Thompson River, are beautiful branches of *Meta-sequoia occidentalis,* the "western dawn redwood".

Despite its name, the first living specimens were discovered in central China in 1944. The McAbee Beds also boast magnificent fossils of cedars, five-needle pines and firs, as well as fish, flies and flowers.

Perhaps most elegant of the Eocene plant fossils is *Ginkgo biloba,* the maidenhair tree. An ancient species, believed by some to be the oldest living tree species, it was already in decline fifty million years ago, according a team of paleontologists who co-authored "Eocene Conifers of the Interior", a chapter in *Life in Stone: A Natural History of British Columbia's Fossils.* As did many other temperate evergreen species, ginkgo disappeared from B.C. (and many other places) when the Earth began to cool. Even in western China, its native habitat, the authors contend, it might not have survived to modern times had it not been treated as sacred and therefore cultivated. Today, the maidenhair tree is back in B.C., as the focal point of residential and public gardens, and on the drugstore shelves as a "the brain herb", a tonic believed by many to improve mild memory impairment and tinnitus, a persistent ringing in the inner ear.

Photos: Peter St. John

The Okanagan Highlands, seen below looking west across the Okanagan Valley above Osoyoos, and (inset) along the shore of Skaha Lake.

The McAbee Fossil Beds

THIS SITE IS UNUSUAL both in its geological and public conceptions. Consisting of sediments laid down near the edge of a shallow lake about 50.2 million years ago, the beds hold magnificently preserved plants, insects, flowers, feathers and fish. Most, aside

from the fish, were apparently blown by the wind onto the lake, before settling to the bottom. But rather than being closed to the public, a 300-metre outcropping of the formation is open to anyone interested in literally digging into the past.

Opening the beds to the public was the inspiration of Dave Langevin, an avid fossil collector who fell in love with the place and eventually purchased the mineral rights. With his partner, Robert Drachuk, they have created a you-dig facility that allows both adults and children to experiment with paleontology and guarantees a fossil to take home—all for a price of course.

Fossils found at the McAbee site (above) include the fan-shaped leaves of ginkgo biloba (top), delicate ferns (at right) and tiny flies. All look remarkably like their modern descendants.

The McAbee Formation, named for the nearby village of McAbee on the Thompson River, is one of the richest and most accessible Eocene deposits in North America. Here is proof that elms grew in Canada fifty million years ago—giving hope that these stately trees might survive the scourge of Dutch elm disease—along with ginkgo and katsura leaves and delicate white pine boughs, all told more than seventy plant varieties. Here too are tiny flies, their transparent wings perfectly preserved, all set into a hillside beneath a wall of much more recent hoodoos.

The site is open from May to October annually; in the shoulder seasons, four-hour tours can be pre-booked; during July and August drop-in visitors are greeted by guides, who provide information, rock hammers and bags. Hard hats and goggles are available on request.

Getting there: The fossil beds are located on the north side of the Trans-Canada Hwy approximately 65 km west of Kamloops.

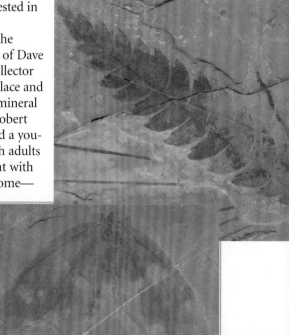

Photos: John Leahy, McAbee guide.

Oregon Jack Creek

The Notch (below) provides a gateway to lush meadows, where mule deer, such as this handsome buck, thrive.

OREGON JACK CREEK is a stream with multiple personalities and a remarkable history. Originating in the highlands of the Cornwall Hills, the creek falls from high limestone cliffs, rushes through a canyon clad with magnificent old-growth Douglas-fir and tumbles through dry brush lands into the Thompson River. That

Peter St. John

Brian Wolitski

confluence likely goes unnoticed by most modern travellers on the Trans-Canada Highway, which passes over the creek about halfway between Cache Creek and Spences Bridge. But early British Columbians knew better. They have camped along the creekside, and on the floodplain and river terraces for at least 8,400 years.

Though to modern eyes, the lower creek might seem a poor spot to camp, the reasons are quickly clear to anyone who turns west on

Hat Creek Road just five kilometres north. Climbing quickly into the hills, the road joins the creek about four kilometres west and follows it up to the limestone canyon and on into the forests and meadows beyond The Notch. Here, over a distance of less than fifteen kilometres, in an area smaller than Greater Vancouver, are rivers and creeks with benches and terraces, hot arid grasslands, cool canyons and valleys, old-growth Englemann spruce forests, alpine tundra meadows, wetlands and aspen parklands.

Each environment has its own flora and fauna. From sturgeon and rainbow trout to rattlesnakes and pocket mice; from mule deer, black bears and cougars to tiger salamanders and spadefoot toads, the Oregon Jack Creek region has it all, including an enormous diversity of birds and a plethora of plants, including mountain potato and nodding onion. And that's without mentioning the salmon, which, 8,000 years ago, were just beginning to repopulate the province's rivers. Little wonder that people made this valley their home for so long.

Archaeologists have also been drawn to this region for almost a half-century. At what they call the Landels Site, they found two layers of occupation below the easily recognizable layer of volcanic ash left by the massive eruption of Mount Mazama in Oregon, an eruption that has been judged to have been forty-two times as powerful as the 1980 eruption of Mount St. Helens. Geologists recently narrowed the timeline for that eruption—which left a enormous cauldera that has been since been filled by Oregon's Crater Lake—to about 7680 BP.

The uppermost layer, immediately below the ash, included microblades, a technology that some archaeologists, including Simon Fraser University's Knut

115

Fladmark, believes originated in Siberia before the end of the last glaciation and travelled across the Bering Land Bridge to Alaska, down the west coast and into the Interior Plateau via the Skeena and Chilcotin Rivers. In addition to the tiny knife blades, there was a deer humerus that proved to be 7,700 years old.

Even more exciting were remnants of an older campsite about twenty centimetres below the first; here, archaeologist Mike Rousseau found microblades and scattered deer bones; the bones proved to be 8,400 years old. Though spear points that are virtually identical to the Scottsbluff, Agate Basin and Eden points used on the Alberta plains more than 9,000 years ago have been found along the middle Fraser and South Thompson Rivers, the Landels Site is the oldest concretely dated site on the Canadian Plateau.

Above the layer of Mazama ash was a small site that appeared to have been a hunting camp. Elk bone allowed the site to be dated to about 5,000 BP. The people who stopped here were big game hunters who also fished for salmon and gathered freshwater mussels.

In 1987, a team of archaeologists from Simon Fraser University did a survey of the Cornwall Hills summit, which lies between the Thompson River and the Upper Hat Creek Valleys. This is an alpine region, with isolated stands of jackpine and Englemann spruce in open alpine meadows about 2000 metres above sea level. The survey turned up nine sites with scattered tools, points and flakes that made it likely that spear points had been manufactured or sharpened. Six of these were on prominent ridges or knolls overlooking alpine meadows or gulleys. Two were on relatively level terrain and contained stone knives (or, as archaeologists term them, "unifacial tools") of the type used for butchering game. The last site, beside a seasonal pond, may have been a hunting camp. Based on the "Kamloops side-notched" arrow heads, three of the sites were dated to between 1,200 and 200 years ago. The others could not be dated.

However, the team did find flakes of basalt that may have belonged to the spear point that spurred the search in the first place. In 1986, archaeologist Pierre Friele was hiking in the hills when he

Cornwall Creek
ELEPHANT HILL PROVINCIAL PARK

Cornwall Hills Provincial Park 2035m

BEDARD ASPEN PROVINCIAL PARK

Three Sisters Creek

Bedard Lake

Cornwall Hills Lookout Forest Service Road

Three Sisters Recreational Site

Hat Creek Road

Oregon Jack Creek

OREGON JACK PROVINCIAL PARK

BLUE EARTH PROVINCIAL PARK

Ashcroft

97c

Thompson River

Basque

Melanie Froese

Lovely long Eden points tipped hunting spears for more than a thousand years, between 9,450 and 8,350 BP.

found the base of a large lanceolate point of medium-grained basalt. Though not whole, the base was very similar to Clovis points, used in many places in North America beginning about 11,500 years ago. Because the point was broken, and because it was a "surface find", it couldn't be dated or even positively associated with the site where it was found. During the survey, however, two basalt flakes, made of the same material as the spear point, were found at the camp-site beside the ephemeral pond, about 750 metres from the point. A small excavation proved only that the site had been used within the past mill-ennium, but perhaps, just perhaps, it had also been used long before.

Later, people all over the Interior Plateau would build villages along the Thompson and Fraser Rivers and their lives would increasingly focus on the annual coming of the salmon. But like the earlier hunters, they would undoubtedly be aware of the riches of upper Oregon Creek, for the canyon and forested cliffs bear witness to generations of native British Columbians who have lived here. Today, within the boundaries of Oregon Jack Provincial Park, which protects the limestone canyon and the falls on Oregon Jack Creek, are culturally modified trees (see page 288) and picto-graphs, indications that this has long been an important ceremonial and spiritual place.

Peter St. John

The view east from the heights of Oregon Jack Provincial Park (above); Agate Basin points (below) were used in many places between 10,450 and 9,450 BP.

Melanie Froese

Getting there: Oregon Jack Provincial Park is accessed by Hat Creek Road, which runs west of the Trans-Canada Hwy 17 km south of Cache Creek. Follow this road, which bisects the park, for approximately 12 km to reach the park. One kilometre farther along, the Three Sisters Forestry Road leads north to the Three Sisters Recreation Site and, about 9 km farther, to Cornwall Hills Provincial Park, which protects a wilderness area featuring old growth Englemann spruce, sub-alpine grasslands and a 360-degree view of the surrounding area. The road, however, can be challenging, particularly after a rain, and the area is home to black bears, cougars, mule deer and moose, among many other animals. Continuing past the Three Sisters junction, Hat Creek Road leads through the park and down from the heights to Hat Creek and the tiny settlement of Upper Hat Creek.

The bare rock of Squally Point looms
above the clear water of Okanagan Lake.

The Okanagan Valley

Mention the Okanagan

and winter escapes, summer fruits
and great Canadian wines spring to mind.
But the Okanagan is much more than these.

Peter St. John

IT'S A PLACE WHERE the history of one of Earth's most remarkable periods of global warming is forever written in the rock; it's also a valley that cradles one of Canada's deepest lakes, where sturgeon so large they were believed to be cousins of the Loch Ness Monster once cruised the depths. In its southern reaches, the Okanagan's arid lands constitute Canada's only true desert, an ecosystem that is—thanks to the sunseekers, orchards and vineyards mentioned above—one of the most endangered in the nation.

This is a valley where the Okanagan people have for thousands of years thrived in summer temperatures that often soared above 40°C, and provided a network of travel and trade routes between the Secwepemc to the north and the Okanogan (as it's spelled on the American side of the border) to the south, and between the Ktunaxa to the east and the Nlaka'pamux to the west.

Millennia after these trails were first beaten smooth, fur traders found the trail along the west side of Okanagan Lake and dubbed it the "Okanagan Brigade Trail", though in honor of their ephemeral fur posts, rather than for the people who had made the land their own since the great glacial lakes filled the valley.

Rather like the Rocky Mountain Trench to the east, the Okanagan Valley is the result of crustal stretching and faulting that began about fifty-five million years ago. From well south of the American border, the Okanagan Valley fault "can be traced north along the valley to a point near Peachland west-southwest of Kelowna," geologists Bill Mathews and Jim Monger write in *Roadside Geology of Southern British Columbia*. "It then takes a jog to the east near Kelowna and … continues just east of the valley as far as Vernon at the north end of Okanagan Lake." Continuing north-northeastward through Sicamous, it is known as the Eagle River fault; in total, the system is about 250 kilometres long. In the Okanagan Valley, it marks the boundary between the Intermontane and Omineca Belts.

As indicated in the section on the Eocene Highlands on page 172, this was a volatile time, geologically speaking. Following the "docking" of the Insular Belt (which includes Vancouver Island, Haida Gwaii and many of the islands off B.C.'s northwest coast), geologists believe there was a change in the direction

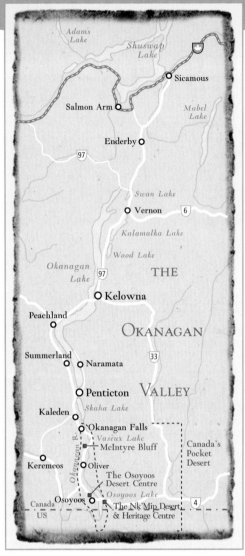

of the movement of the Pacific Plate and other smaller plates off North America's west coast. Instead of slamming head-on into and under the North American Plate, they began sliding past one another. The result: crustal stretching, violent faulting and a line of volcanoes that stretched from B.C.'s Interior Plateau south into the U.S.

In the last two million years, repeated glaciations have rounded the uplands and scoured the valleys, making them perfect conduits for the meltwater that inevitably followed and for the many species of fish eager to recolonize the ice-free lakes and rivers.

At the end of the last glaciation, the Okanagan Valley served as a major outlet for the water of Glacial Lake Thompson and Glacial Lake Deadman (see page 104) that filled what are now the South Thompson and Thompson River Valleys. The water poured south into the Columbia River until, between 10,450 and 12,500 BP, the ice dam at the western end of Glacial Lake Deadman finally gave way, draining it in a catastrophic flood that drilled through the Fraser

Canyon, thundered down the Fraser Valley and raced across the Strait of Georgia.

Cut off from its major source of water, Okanagan Lake and its smaller sisters—lovely Lake Kalamalka to the east and Skaha, Vaseux and Osoyoos Lake to the south—were sustained by a watershed that stretched from the Monashee Mountains on the east to the Okanagan Highlands on the west. In fact, thanks to the Okanagan fault, the lake has a maximum depth of 244 metres or 760 feet. Okanagan, by the way, means "place of water."

The lake and some of the streams that feed it, including Deep Creek just south of Peachland, are known for their significant runs of spawning kokanee salmon, but Okanagan Lake also has a unique shore spawning population.

The Okanagan River still flows south to meet the Coumbia, as it has for millions of years, but for at least one distinct population of salmon— the Okanagan Valley chinook— coming home has become more and more of a challenge. Okanagan chinook is the only remaining chinook population in Canada that migrates to the Pacific through the United States via the Columbia River. Their numbers have dwindled as habitat loss, historic overfishing, and the gauntlet of Columbia River dams have made it ever more difficult for returning salmon to reach their spawning grounds. In 2005, fewer than fifty fish returned to spawn, spurring calls to put Okanagan chinook on Canada's endangered species list.

Kokanee Salmon

KOKANEE (which is Ktunaxa for "red fish") salmon are almost indistinguishable from their first cousins, the sockeye, except for one significant difference: kokanee live their entire lives in fresh water, while sockeye make a remarkable journey to the ocean, spending the majority of their lives in the Pacific before returning to spawn.

There have been several explanations for the great environmental (or is it a cultural?) divide between the kokanee and the sockeye, including one that has kokanee yearlings poised, en mass, at the top of Bonnington Falls (see page 73), peeking over the precipitous edge and, realizing they would never be able to return to the streams of their birth, making a joint decision to become freshwater fish.

This is a rather unlikely portrayal. But there is a more credible explanation. Following the end of the last glaciation—and indeed, in the midst of it, as the fossil kokanee of Kamloops Lake (see page 108) have demonstrated—the ice melted to form great glacial lakes. These waterways allowed the Okanagan, Kootenay and Thompson River systems to be populated with salmon and many other species of fish via the Columbia River, the mouth of which lay south of the ice. Over the millennia that these river systems were connected, a separate, freshwater population of sockeye evolved, as it also did in the continental U.S., Alaska, Siberia and Japan. By 9,500 BP, when the glacial lakes had disappeared and water levels had dropped, waterfalls such as Okanagan and Bonnington Falls

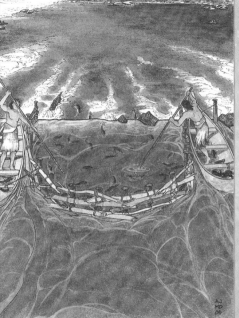

became upstream barriers for the anadromous, or ocean-going, sockeye salmon in the Columbia River. In the Okanagan and elsewhere, their absence meant that the landlocked kokanee (or as they're known in the Okanagan, the kickininee, little red-fish, silver trout or yank) were left to spawn on their own. In other river systems, including tributaries of the Columbia and Thompson River and the Shuswap Lakes, spawning sites were shared with anadromous sockeye that had made the long upstream trip from the ocean.

Interestingly, however, recent genetic testing of various kokanee populations have shown that the kokanee of the Okanagan are more closely related to the sockeye of the Columbia River than they are to the kokanee in Kootenay Lake. This implies that each lake's freshwater salmon population evolved for its own specific conditions, and is therefore uniquely adapted to its home waters.

Okanagan Lake supports two types of kokanee salmon— stream spawners and lakeshore or beach

Heading upstream, salmon leap over a low falls to find themselves trapped in a weir of woven logs and branches. As they seek a way through the barrier, they are speared by fishermen standing in canoes tied to posts anchored in the riverbed.

Amanda Dow

Deep Creek, with lovely Hardy Falls, is not only a popular place for walking and picnicking, even for grizzly bears in years past, but one of the region's key spawning streams.

Peter St. John

Dennis Fast

this last crucial act of their lives. And DNA studies have shown that the stream spawners are genetically distinct from the beach spawners. Remarkably, these two types of kokanee are only found in the Okanagan—in Okanagan, Kalamalka and Wood Lakes.

In appearance, kokanee are much like their salt water cousins, though smaller, averaging less than two kilos in weight. Through most of their lives, they are silver in color, turning bright red, like the sockeye, as they spawn. Salmon fry and yearlings are actively hunted by trout. This makes life particularly onerous for the kokanee of Kootenay Lake, for there the salmon fry and yearlings must contend with the remarkable Gerrard trout, a subspecies of rainbow trout that can grow to weights of twenty-four kilos or more than fifty pounds.

Nor is life is easy for the salmon young in the Okanagan. Because the lake is oligotrophic —poor in plant life and nutrients, though rich in oxygen— many tiny salmon are not able to get enough to eat to survive their first winter.

Further, some of the best known spawning streams in the Okanagan Valley, including Kelowna's Mission Creek and Peachland's Trepanier and Deep Creeks are under pressure from the region's rampant development. However, there is a growing awareness about the importance of kokanee salmon to the Okanagan in terms of the valley's heritage and culture, as well as its tourism value.

spawners. The two types mix in the lake during their entire lives, but when the spawning urge comes upon them—generally when they are three years of age, though some fish do not spawn until they are four or even five—the two groups separate. In early September, the stream spawners migrate into the lake's fourteen major tributaries to lay their eggs, a process that is complete by mid-October. The beach spawners, meanwhile, wait until mid-October to begin spawning along the beaches of the lake's long shoreline. Tagging fry of both populations has demonstrated that they are completely separate in

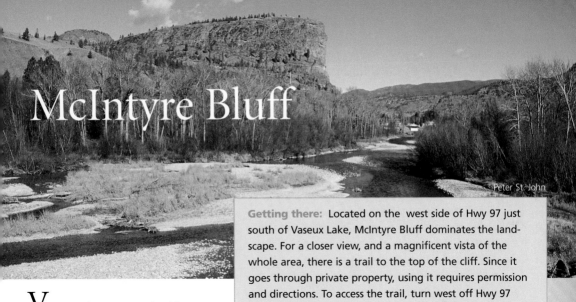

McIntyre Bluff

Peter St. John

Getting there: Located on the west side of Hwy 97 just south of Vaseux Lake, McIntyre Bluff dominates the landscape. For a closer view, and a magnificent vista of the whole area, there is a trail to the top of the cliff. Since it goes through private property, using it requires permission and directions. To access the trail, turn west off Hwy 97 onto Seacrest Road (386 Ave.) immediately after crossing the bridge over the Okanagan River just south of the bluff. Wind your way uphill and turn right into Covert Farms. Stop at the office for permission, parking information and directions. The hike takes approximately four hours, but the view from the summit is worth the effort. A new trail organization in the region is planning hikes on the Kettle Valley Railway Trail from Okanagan Falls to Osoyoos, with an ancillary route to McIntyre Bluff.

For those wishing to visit the new grasslands park, head south along the Green Lake Road from Hwy 97 at Okanagan Falls.

VISIBLE from a considerable distance, McIntyre Bluff, on the west side of the Okanagan Valley between Okanagan Falls and Oliver, is one of the few places west of the Rockies where the basement of North America can be clearly seen. This 667-metre cliff of Precambrian gneiss, known as the Vaseux Formation for the lake just to the north, it is a spectacular testament to the power of both tectonic plates and glaciers.

Buried for hundreds of millions of years beneath more recent layers of North America's ancient margin, geologists believe this massive block of granite was revealed during the Eocene, about fifty million years ago, when a large piece of Quesnellia, part of the Intermontane Superterrane, slid west, opening an enormous, west-dipping fault. Much later, valley sediments, as well as the cliff face, were scoured clean by the glaciers that have pushed through the valley more than once over the past two million years.

The cliff sits at the narrowest part of the Okanagan Valley, a bottle-neck that was also the site of an ice dam instrumental in creating Glacial Lake Okanagan at the end of the last glaciation. Today, the Okanagan River flows past the base of the cliff.

Similar rock can be seen on the other side of Vaseux (which is French for "Muddy") Lake.

Just northwest of McIntyre Bluff, a large tract of mid-elevation grasslands, with old-growth ponderosa pine stands, has recently been put under provincial government protection. Zigging and zagging around vineyards and farms, the park stretches north from the cliff along the west side of Vaseux Lake to Mount McLellan and Mount Hawthorne, and west to Green Lake and the community of Willowbrook. Not only is the area, which is known as White Lake Grasslands Provincial Park, an important wintering and lambing range for bighorn sheep, it also contains fifteen endangered or threatened bird species, including the gravely endangered whiteheaded woodpecker, as well as red- or blue-listed species of reptiles, mammals, amphibians, insects and plants.

Not surprisingly, such an abundant environment was well used by the Okanagan or Syilx people; recent surveys have found eight upland archaeological sites, as well as rock shelters and pictographs.

Viewed from the south, McIntyre Bluff, above, dominates the Okanagan River Valley north of Oliver.

Canada's Pocket Desert

WITH AN ANNUAL PRECIPITATION of about twenty-eight centimetres (or eleven inches) the shrub-steppe ecosystem of the South Okanagan Valley is Canada's only true desert. Some have called this rapidly disappearing ecosystem, which stretches along the valley south of Skaha Lake, the northern tip of the Sonoran Desert. In fact, the Sonoran Desert, which is located far to the south, has quite rigidly defined boundaries that extend from south-central California through the southern half of Arizona east almost to New Mexico. Instead, both the shrub-steppe ecosystem and the Sonoran Desert are part of a vast arid region between the Rocky Mountains on the east and the

Big sagebrush (above), ponderosa pines (above and opposite page) and bighorn sheep (opposite) can all be found in what has been called "Canada's only desert".

Cascade and Sierra Nevada ranges to the west. Canada's "pocket desert", as the shrub-steppe ecosystem has been called, is the northernmost tip of this continental region. Like the Great Basin Desert of Idaho, Utah and Nevada, the shrub-steppe ecosystem (known by locals as the "antelope-brush desert") is a cold desert, where precipitation falls as snow in the winter and the dominant plant life is not subtropical. Yet dwellers of more southerly deserts would recognize at least some of the vegetation here. Big sagebrush, for example, can be found farther south, and is one of the few native species that has been assisted by settlement and the accompanying overgrazing of livestock. Evergreen and remarkably hardy, this aromatic plant, which is also known as big sage, common wormwood and basin sagebrush, grows in vast tracts in the U.S., covering 470,000 square miles across eleven western states. Nevada has big sagebrush as its state plant.

Growing up to 1.2 metres tall, this branching gray-green shrub was used by the Okanagan people in a wide variety of ways: the wood was burned for fuel and the leaves, which contain camphor, were used to make teas to treat colds and coughs, and to soak sore feet. Branches were widely used as a fumigant and to the northwest, the Nlaka'pamux used the bark, which naturally shreds, to weave mats, bags and cloaks.

Antelope-brush, the fragrant shrub that gives Canada's desert ecosystem its nickname, is often found growing with big sagebrush in dry, sandy grasslands soil. It's easy to spot in late spring, with its shaggy, upright branches and bright yellow flowers, and provides crucial food for deer and California bighorn sheep. Farther south, it helps sustain North America's gravely endangered desert pronghorns.

Peter St. John

Jerry Kautz

tall grass with smooth, stiff gray-green leaves that were widely used to decorate baskets and by Okanagan women to line steam pits and food caches.

Peter St. John

Antelope-brush is also known as antelope-bush, bitter brush and greasewood. The last of these refers to the plant's pitchy quality that, according to ethnobotanist Nancy Turner of the University of Victoria, "makes it good for producing a hot fire quickly". Bundles of branches served the Syilx or Okanagan people as a dependable and portable fire starter that was often used in winter or when travelling.

The seeds, which are shaped like tiny spindles, were widely eaten by chipmunks, grounds squirrels and deer mice; seeds that had been cached for the winter in underground burrows were often overlooked, resulting in clumps of antelope-brush seedlings in the spring.

Another shrub found here, as well as in the Thompson and Fraser basins and the East Kootenays, is rabbit-brush, a smaller shrub with gray, velvety leaves that served Okanagan and Secwepemc women well during their childbearing years: the leaves were used as sanitary napkins and to make a tea to ease cramps. True to its name, its leaves and stems are a favored food of jackrabbits, as well as deer and mountain sheep.

This arid environment also sustains milkweed, the monarch butterfly host, and giant wild-rye, a

Saskatoon bushes grow in many places along the Okanagan Valley. The fruit—called "real berries" by the Secwepemc—was "the most popular and widely used berry for central and southern native peoples", according to the multi-authored guide *Plants of Southern Interior British Columbia and the Inland Northwest*. Certainly residents of the prairie provinces, past and present, would agree; many deem saskatoons far superior in taste to the more commercially available blueberries. In the southern Okanagan, the berries were collected in great numbers, then dried and traded west to coastal peoples.

On the benches and valley slopes, ponderosa pines grew to heights of thirty metres or more, forming beautiful open forests that provided winter range for deer, elk and mountain sheep, seeds for many small animals, and crucial nesting and foraging areas for many birds, including the endangered white-headed woodpecker. Unfortunately, the wood of the ponderosa pine (which is also known as yellow pine, western yellow pine, bull pine and rock pine), proved valuable for use in construction and very little of this magnificent old-growth forest remains. A glimpse of it can be seen at the Nk'Mip Desert Cultural Centre just northwest of Osoyoos (see page 128).

The Osoyoos Desert Centre

ESTABLISHED IN 2000 to protect British Columbia's rapidly disappearing antelope-brush desert, the Osoyoos Desert Centre north of Osoyoos is one of two sites in the South Okanagan that is actively engaged in research, restoration and public education about this remarkable environment. The other, Nk'Mip (see page 128), on the east side of Osoyoos Lake, blends information about the desert with a fascinating glimpse of the people who have lived here for thousands of years.

Not only is this Canada's only true desert, it is also one of the nation's four most endangered ecosystems; the others are Vancouver Island's Garry oak woodlands, Manitoba's tall grass prairie and southern Ontario's Carolinian forest. While the Okanagan people have long appreciated its true value, until recently, many others have dismissed it as "wasteland" and transformed it into orchards, vineyards, golf courses and suburbs. Today, experts estimate that about thirty-five per cent of the antelope-brush ecosystem remains. But now, perhaps just in time, its merits are finally being realized.

Fully thirty per cent of British Columbia's wildlife species at risk (a total of eighty-eight species) make the southern Okanagan and Similkameen Valleys their home, and twenty-two per cent of all threatened and endangered vertebrates as well as hundreds of rare insect and plant species can be found here. Among those are Canada's only praying mantis; small populations of endangered (or red-listed) burrowing owls, northern leopard frogs and white-headed woodpeckers; rare, nocturnal pallid bats; four types of snakes including the western rattlesnake, a threatened (or blue-listed) subspecies of wolverine, and bighorn sheep, which are under pressure everywhere in their desert range.

Photos: Peter St. John

For aspiring botanists, this is a fascinating eco-system, with fragile lichens and mosses that conserve what little moisture the area receives, prickly-pear cacti with their edible fruits and nec-tar, and magnificent flowers, including the rare bitterroot and threatened mariposa lilies, which provide bril-liant counterpoints to the perennial bunchgrasses, sage-brush, antelope-brush and rabbit-bush. On the hills above the valley, stately ponderosa pines grow in widely spaced stands. All these plants have long been used by the Okanagan people.

The Osoyoos Desert Centre was created not only to protect this fragile ecosystem and the many species, such as badgers, tiger salamanders, sage thrashers and night snakes, that face being extirpated, but also to increase awareness about the intrinsic value of the desert. Clearly, as lovers of the tall grass prairie also know well,

Peter St. John

marshalling support for the preservation of an old-growth grasslands environment is con-siderably more difficult than creating appreciation for a majestic old-growth forest.

Nevertheless, in a short time, the Osoyoos Desert Society, which runs the centre, has gone a long way toward creating awareness about this intensely threatened ecosystem. Its full program includes research, restoration and education, but vis-itors are encouraged first to experience this remark-able ecosystem through guided, self-guided or virtual tours. A 1.5-kilometre system of elevated boardwalk trails protects the fragile environment, while inter-pretive kiosks, signs and pamphlets provide informa-tion about this unique "pocket desert".

Members of the society play an important role in expanding the centre's increasingly varied activities, and many are also passionately involved in petitioning the government of Canada to establish a national park reserve in the South Okanagan and Similkameen Valleys (technically B.C.'s Interior Dry Plateau), one of Canada's thirty-nine terrestrial regions that has yet to be represented by a national park.

Getting there: The Osoyoos Desert Centre can be reached by heading north from Osoyoos on Hwy 97 toward Kelowna. About five km north of the city, turn west (or left) on 146th Street at the sign for the Desert Centre. Continue up the hill until you see the centre on the south side of the road. The centre opens each year on Earth Day in April; for tour times, call 1-877-899-0897 or email mail@desert.org.

To Kelowna

Osoyoos

To Vancouver

Turn west on 145th Ave

Osoyoos Desert Centre

Lake

3

97

To Calgary

OSOYOOS

CANADA / US

The Nk'Mip Desert Cultural Centre

CRADLED BETWEEN the Okanagan Highlands and Osoyoos Lake, Nk'Mip, the dramatic lake-front desert reserve of the Osoyoos people, recalls the Roman god Janus. With two faces, this god of change and transition was able to look back and forward at the same time, a balance of past and future that

they have also created a magnificent heritage centre that celebrates the long history of the Syilx or Okanagan people in their desert environment.

Three kilometres of wide, well-groomed trails wind through sage-

Wide paths wind through the open ponderosa pine forest, leading to a reconstructed village, including a sweatlodge (inset).

Photos: Peter St. John

heralded a golden age in Roman mythology. The same might be said of the Osoyoos people who, in less than two decades, have transformed Nk'Mip—their "bottom land"—into one of British Columbia's most remarkable destinations.

Led by their visionary chief, Clarence Louie, the Osoyoos people have developed an award-winning winery, a luxury resort with a wine-tasting facility that puts many in Europe to shame, a full-service spa, a lake-front campground and a nine-hole golf course. But most important as far as this guide is concerned,

brush grasslands and meander among huge ponderosa pines. This, like the Osoyoos Desert Centre on the other side of the lake, is part of

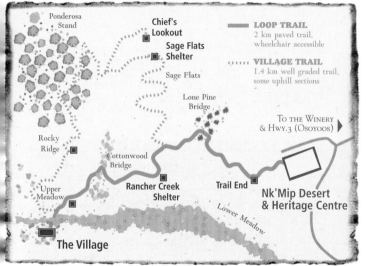

The Village

- Ponderosa Stand
- Chief's Lookout
- Sage Flats Shelter
- Sage Flats
- Lone Pine Bridge
- Rocky Ridge
- Cottonwood Bridge
- Rancher Creek Shelter
- Upper Meadow
- Lower Meadow
- Trail End
- To the Winery & Hwy.3 (Osoyoos)
- Nk'Mip Desert & Heritage Centre

LOOP TRAIL
2 km paved trail, wheelchair accessible

VILLAGE TRAIL
1.4 km well graded trail, some uphill sections

Getting there: The Nk'Mip Desert Cultural Centre is located on Hwy #3, one km east of Osoyoos. Admission is charged at the heritage centre and annual passes can be purchased. The main trail is wheelchair accessible and both it and the longer loop trail have shelters and benches where visitors can rest. Because temperatures can soar during spring and summer, hats and sunscreen should be worn and water should be carried.

Canada's only real desert. In fact, Nk'Mip and other reserves north of it are largely responsible for saving one of the nation's most endangered ecosystems. Almost sixty per cent of what remains of this unique landscape is on native land.

Winding beneath ancient ponderosa pines—some are several hundred years old—the path leads to a reconstructed Okanagan village. Here, a tule-mat tipi, a sweatlodge and two kekulis, or pithouses, allow visitors a real glimpse into the lifestyles of the ancestors of these resourceful people.

Dug deep into the earth, with a thick roof of logs and sod, each huge pithouse is cool and quiet even when the temperature soars to 35°C or more. But wait. The room, already shadowed, turns suddenly dark as a cloud blocks the sun. Now the spring air is filled the smell of rain and a wind, sweeping through the side entry, brings the sweet scent of sagebrush. Thunder cracks overhead, the skies open and suddenly it's easy to imagine a long ago spring storm and possible, almost, to hear the echoes of the people who once lived in lodges like this.

Then our guide, Brenda Baptiste, laughs and the vision vanishes in a rush of modernity, for two of the centre's employees are pelting up the broad path with umbrellas.

Back at the beginning of the trail, the magnificent new Nk'Mip Desert Cultural Centre broadens these experiences, using sculpture, exhibits, films and state-of-the-art technology to impart the stories and lessons of the past.

Janus would feel right at home.

Dug deep into the earth, kekulis or pithouses stayed cool during the summer months and, warmed by a fire, were cozy during the winter. The Okanagan people also used light mat lodges, constructed of bulrushes sewn together with nettle or hemp fibres, during the summer. For travel on the region's lakes, large cottonwoods were hollowed out to create canoes, opposite top.

Photos: Peter St. John

The Western Rattlesnake

While it's wise to keep one's distance, rattlesnakes are perhaps more misunderstood than malicious.

Dennis Fast

BEGINNING IN 2002, biologists and a group of apprentices began a project at Nk'Mip to learn more about British Columbia's only venomous snake, *Crotalus oreganus*, the western rattlesnake. Confined to the dry valleys of the southern Interior, this large snake is in decline, for several reasons. Its fearsome reputation has led to a needless slaughter; a growing number of highways has both killed snakes and cut them off from their grassland feeding grounds, and agricultural and urban development have resulted in destruction of their dens and foraging areas. The result is that western rattlesnakes are now blue-listed and considered vulnerable in the province.

The Nk'Mip study has several objectives. The first is to learn about the population on the reserve. Individual snakes are captured, weighed, measured, tagged with tiny tags that will allow them to be easily identified, then released back into the desert. A few snakes have been equipped with radio-transmitters,

enabling researchers to study their movements. Using the information thus gathered, the study then hopes to raise awareness about rattlesnakes in general and ultimately to reduce the threats to these fascinating animals and their habitats.

To minimize the likelihood of unexpected encounters that might prove dangerous for both snakes and humans, Nk'Mip was developed with wide, well-kept footpaths that allow snakes attempting to cross to see oncoming human traffic, and humans the necessary time and distance to allow mobile snakes to go about their business. Thus far, both parties seem to have found the design to be satisfactory.

Found in hot, dry grassland habitats in rocky terrain, western rattlers are at the northern end of their range in B.C. They hibernate from September to April, returning each year to the same communal den, which, rather suprisingly, they sometimes share with other species, including garter snakes. Generally located in caves or rocky fissures on south-facing slopes or ridges, the dens are located far enough underground for the winter temperature to remain above freezing. Emerging in the spring, each snake will travel up to a kilometre to its favorite foraging and basking area, where it spends the long summer days hunting rodents in the grasslands and basking on sunny rock ledges, often close to shrubs that provide shade.

Rattlesnakes are found in a number of places in the South Okanagan. The sign on this page was posted at the Osoyoos Desert Centre (see page 126) on the west side of Osoyoos Lake.

Mating takes place in late summer and the females store the sperm over the winter before ovulating and fertilizing the sperm in the spring. According to studies done by the B.C. government, pregnant females do not feed and stay close to the wintering den during the summer months. In September or early October, more than a year after the mating has taken place, between two and eight live young are delivered. The newborns immediately begin to feed and grow rapidly, shedding their skins once before joining the rest of the population in the den; despite their caloric intake, about a quarter will die before spring. Their mother, however, begins hibernation immediately and will not feed until the following April, a fasting period of nineteen or twenty months, which rivals that of pregnant and nursing polar bears.

Because of this arduous pregnancy, females do not begin to breed until they are between seven and nine years of age and then reproduce only every three years on average; following birth and hibernation they must double their weight before attempting to breed again—one of nature's most extreme examples of yoyo dieting.

Adults range between two-thirds of a metre and 1.5 metres in length and have been known to live up to twenty-five years, but their slow reproductive rate and an increasingly fragmented habitat means that small populations are particularly vulnerable.

Despite their reputation, rattlesnakes are shy animals and will go to considerable lengths to avoid confrontations. When approached, they will almost always seek cover and often give a distinctive warning rattle. If you see or hear a rattlesnake, freeze and back away, giving the snake room to retreat. Rattlesnakes are protected under the B.C. Wildlife Act and killing or capturing one for other than authorized scientific study is prohibited.

To learn much more about rattlesnakes, visit the Nk'Mip Desert Cultural Centre.

Peter St. John

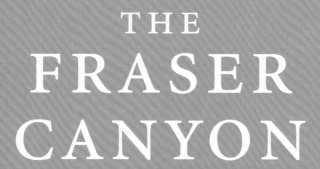

THE
FRASER
CANYON

The beautiful Stein River tumbles through its
pristine valley toward its meeting with the Fraser.

Dennis Fast

Between Lillooet and Pavilion, the Fraser Canyon is much broader than it is downstream, providing evidence of its greater antiquity.

The Fraser Canyon

The spectacular

mid-section of the Fraser River

follows more than a half-dozen fault lines

that run northwest to southeast between Big Bar and Yale.

GEOLOGISTS BELIEVE the long and complicated history of this area is likely related to both the uplift of the Coast Mountains to the west and the rising of the Interior Plateau to the east. This is particularly true of the broader northern section of the canyon north of Lillooet, which may be as much as twenty million years old. The more celebrated southern portion, the almost dizzyingly steep and narrow canyon bordered by the Trans-Canada Highway south of Lytton, seems to indicate a much shorter history.

Is this, as geologists John Clague of Simon Fraser University and Bob Turner of the Geological Survey of Canada surmise, the work of a piratical river aided by roaring meltwater from the repeated glaciations of the Pleistocene—the great Ice Age—over the past two million years? The combination of rounded mountains above and precipitous walls below seems to indicate a two-part process: glacial scouring followed by the

drilling and cleaving action of the meltwater that inevitably followed each glaciation. From Lytton south to Yale, the river gradient drops more than ninety metres or 295 feet.

Though breathtakingly picturesque, throughout its length the canyon (which today can be viewed from campgrounds, rest stops, an air tram and the road itself) is in places all but impassable. At Hell's Gate, near Boston Bar, its walls narrow to barely thirty-five metres apart and rise nearly 1,000 metres above the thundering water. Farther upstream, just north of today's Lillooet, rocky ledges choke the river near its confluence with the Bridge River. Yet these apparently inhospitable spots were for millennia among the most popular gathering places for early British Columbians. At certain times of the year, it is believed that crowds of more than 10,000, including people from hundreds of kilometres away, assembled here. The reason? Salmon, of course.

The best fishing sites were the places where the river's narrows or natural barriers formed obstacles to the innumerable migrating salmon. And since during the past 5,000 or 6,000 years salmon have provided the majority of the protein and fat for many cultures, the best fishing sites drew the greatest crowds. It was salmon, too, that made the lower Fraser, just below the mouth of the canyon, home to a greater population

Peter St. John

The granite ridge that long protected one of British Columbia's oldest summer encampments can be clearly seen from across the Fraser River, just north of Yale.

extending fifteen metres or nearly fifty feet into the sandy soil. Seven of these lay below a clearly defined layer of Mazama ash, from the catastrophic explosion of Mount Mazama almost 7,700 years ago. Among the many features discovered, in addition to five hearths, many postholes and a rock wall, were dozens of beautifully crafted spear points, as well as micro-blades, choppers and pounding stones.

The spear points, along with chokecherry pits and even the soil itself help to tell the story of the site, and by extension, several chapters in the long tale of the river. Layers of coarse gravel and stones at the base of the campsite indicate that when the canyon was younger—and the water higher —the river flowed right through the site. Later, as the Fraser began to cut a course to the west, the site became a high water bed. Until about 6,000 years ago, when the high water level of the river dropped below the terrace, silt and sand were deposited almost every spring in the bowl-shaped area behind the ridge. The result made it unsuitable as an encampment until the soil dried in midsummer, but by late summer the site was perfect. So time and again the people returned just as the chokecherries began to ripen and the first sockeye salmon began to migrate upstream.

Standing next to a narrow channel below the terrace, Stó:lo fishers may have been able to spear migrating fish as early as 7,500 years

density than almost anywhere else in ancient North America. Where the banks of the river permitted settlement, campsite after campsite lined the water's edge, or perched above the river.

The oldest of these sites, including a beautiful but largely inaccessible location on the east side of the river north of Yale, just upstream from Siwash Creek, points to the early post-glacial presence of salmon runs. Archaeologists call this the Milliken Site, after local settler August Milliken. But the Stó:lo people, who have been here far longer, call it Aselaw. By whatever name, this is a place where people have lived for almost 10,000 years.

Perfectly situated on a small alluvial terrace behind a protective granite ridge, the site proved to have ten "cultural zones", as archaeologists term them, ago, when biologists believe salmon stocks began to repopulate the northern rivers in significant numbers. Salmon were an eagerly anticipated addition to diets that, in the early millennia after the ice melted, relied mainly on lean meat and plants.

People living in the Interior Plateau continued to be mainly hunters rather than fishers for much of the next 3,000 years, but as the centuries passed, they added freshwater clams and plants of many kinds to their diets.

Their weapons and tools reflected this dietary variety. Well made and regularly sharpened, their spear points were smaller than those of their predecessors, often with notched corners and barbs. Reflecting trade or contact with the West Coast, they also used microblades—tiny razor-like knives—as well as bone points and needles and antler wedges for woodworking. Archaeologists call this adaptable lifestyle the Nesikep Tradition, after a creek that feeds into the Fraser from the west, south of Lillooet.

Over time, as the fish multiplied and returned with ever more dependable regularity, the descendants of these early fishers, as well as many other peoples throughout the Fraser River Valley and its many tributaries, increasingly organized their lives around the coming of the salmon.

By 3,000 BP, the salmon had become the heart of more than a half-dozen societies, the currency of their cultures and the key to their identities.

Though the Milliken Site is currently inaccessible except by water, the Stó:lo people in the surrounding area hope to organize tours of this fascinating ancient place.

Other, somewhat less venerable sites along the lower Fraser River Canyon include Xwoxewla:lhp, "Willow Trees", the South Yale site. Located opposite the town of Yale, it has been more or less continuously occupied, at least seasonally, for an estimated 6,000 years. One can still find Stó:lo fishers there today.

And Saddle Rock, just south of the Saddle Rock Tunnel about five kilometres north of Yale, marks the first major set of rapids as one travels upstream. It is still

Lady Franklin Rock, (inset), is one of the largest boulders from a landslide that blocked the Fraser River about 7,200 years ago. The slide scar is easy to see on the river's east side.

an important fishing site, though there are no safe pullouts at which to park.

Images this page courtesy of Roy Carlson

137

Just above Yale, what is locally known as Lady Franklin Rock marks the site of a huge landslide that geologists have recently dated at about 7,200 BP. Thundering down from the east side of the river, a huge mass of rock (including the aforementioned "Lady") slid into the Fraser and dammed it. When the water broke through, according to archaeologist Roy Carlson, "the flood caused the river channel high up [on the terrace] at the South Yale site farther downstream."

Farther up the Fraser, along the banks of Keatley Creek north of Lillooet, archaeological teams led by Brian Hayden conducted digs for years, eventually uncovering a large, complex, stratified town that thrived for centuries on what appears to be a lacustrine or lake terrace created by the glacial lake that briefly filled the valley following the last glaciation. The story of Keatley Creek can be found on page 143.

Not surprisingly, perhaps, given the steep canyon walls and the surrounding mountains, rock slides have been and continue to be a threat to both humans and salmon. Brian Hayden believes that a devastating rock slide about 1,100 years ago may have forced the abandonment of Keatley Creek, as well as several other large communities along the middle Fraser River. Blocking the migration of the salmon,

and thereby not only killing them in enormous numbers but preventing their spawning, could have had such a cataclysmic effect on the Fraser River salmon population that it spelled the end of a sophisticated culture that had occupied the valley for nearly 1,500 years.

Other landslides, including the largest known rock avalanche in Western Canada, have impacted the lives of B.C. residents to varying degrees for millennia (see page 161).

Other canyons along the river include the somewhat grandiosely named Grand Canyon of the Fraser, a short gorge in the Rocky Mountain Trench, far upstream and east of Prince George. No one who has seen both would confuse it with the Fraser Canyon. And many of the dozens of creeks and rivers that feed the Fraser south of Williams Lake spill through canyons of their own; they include Farwell Canyon on the Chilcotin River, the arresting Marble Canyon (see page 141), the magnificent Upper and Lower Canyons of the Stein River (see page 146) and the canyons of the Coquihalla River, which enters the Fraser near Hope.

Lil'Wat fishermen, shown here in 1862, and other Fraser Canyon peoples shared the river's incomparable wealth with European settlers who began arriving in ever greater numbers at the end of the nineteenth century.

J.R. Mackey / National Archives of Canada / ISN-103055

Wool Dogs

DOGS HAVE PLAYED an important part in the lives of people worldwide for millennia. They have served as herd animals, hunters, guards and sentinels; they have pulled travois on the plains and sleds in the far north, and many breeds have served as loyal companions. But perhaps the world's most unusual canine breed was developed by Salishan-speaking peoples of what are now southwestern B.C. and neighboring Washington State.

hair, this remarkable breed, now unfortunately extinct, was known as the wool dog: *sko-mai* or *ki-mia* in the Coast Salish languages, and *xlit selken* among the Nlaka'pamux of the Thompson River region, who speak an Interior Salish dialect. Like sheep or goats that are raised for their fine fleece, wool dogs were carefully bred and isolated, often on islands or in pens, from other breeds. And like their ruminant

Just in time, artist Paul Kane immortalized, in the romantic style of the time, both a wool dog and a blanket being created following one of two journeys he took across North America in the mid-1840s.

Royal Ontario Museum, 2005-5163-1

Medium-sized, with thick, long, soft, generally white or dun-colored

counterparts, they were regularly shorn close to the skin. But their wool was not used for just any type of garment; instead, sometimes combined with the fine wool of elusive mountain goats, as well as plant fibres, it was used to create the magnificent blankets for which coastal British Columbians are celebrated even today.

For hundreds of years, these beautiful blankets

were perhaps the ultimate symbol of wealth, given as gifts at potlaches, where they would best display the donor's generosity and affluence, and used to wrap the dead for burial.

Despite the burial practice, the blankets themselves are rarely found in archaeological sites, since most natural fibres rapidly decompose. As a result, it's not clear when they were first made and, by extension, when wool dogs were first successfully bred. Archaeologist David Burley has suggested that, since breeding for such specialized traits would take time and patience, and since the resulting wool and blankets would only be of value to a complex and highly stratified culture, wool dogs may well stretch back to the early Marpole culture some 2,500 years ago or even to the Locarno Beach culture that preceded it during the millennium before.

Several pieces of evidence combine to suggest that wool dogs were a part of life for people from Vancouver Island to the Fraser Canyon and south to the Olympic Peninsula by 1,500 BP. Victoria archaeologist Grant Keddie believes that two small dogs buried in middens dating from that period on southern Vancouver Island may be wool dogs. In part this is because he has also discovered, in midden layers dating to the same period, blanket pins, spindle whorls and antler combs (the last of these possibly used for tamping the weft or woven fibres tightly to create a sturdy woven cloth).

Linda Fairfield

Roy Carlson of Simon Fraser University, and others, have frequently found decorated and highly

polished bone and antler blanket pins, as well as spindle whorls of whalebone or a soft, easily carved soapstone called steatite in late Marpole sites on the Lower Mainland. Virtually all are 1,500 years old or younger. Older spindle whorls are rarely found, likely because most were made of wood.

Wool dogs clearly continued to be bred and their wool continued to be used ceremonially for more than a millennium, for many of the first European adventurers and traders, including George Vancouver, Simon Fraser and Alexander Mackenzie all made a point of commenting on both the animals and the remarkable end products. Of his visit to a village near present-day Yale in 1808, Fraser wrote, "They have rugs made from the wool of Aspai or wild goat, and from Dog's hair … we also observed that the dogs were recently shorn." Villages in the area were apparently famous for their exquisite blankets or "rugs".

As it was for so much else, the arrival of Europeans was the death knell for the remarkable breed. Traders soon flooded the market with cheap blankets and, later, sheep's wool. And without careful breeding, the dogs quickly disappeared. By 1858, wrote historian F.W. Howay, "the blankets themselves were very scarce, and the wonderful dog had become almost extinct." Writing in the *Canadian Journal of Archaeology* in 1994, author Rick Schulting wrote, "The last account of what was possibly a living pure-bred wool dog comes from 1862, when the first postmaster of Vancouver seems to have seen a dog fitting that description at a Lower Fraser band potlatch."

Marble Canyon

Peter St. John

LOCATED BETWEEN the Fraser and Thompson Rivers, this stunning little canyon with its emerald lakes and rugged chalk-faced slopes is more than just a scenic spot to stop for lunch. Here, visitors can see the result of some of British Columbia's most complex tectonic plate movements and also find, in the limestone cliffs around Pavilion Lake, one of the very few places in the world where the fossilized remains of freshwater stromatolites—some of Earth's oldest lifeforms—can be seen.

Like the fault-prone area around Hell's Gate, Marble Canyon was once part of a chain of Pacific Islands, lying far to the southwest of the coast of ancient B.C. Over time, it rode north and was eventually caught between the westward moving North American plate and the edge of the Pacific Plate, which was moving east. On the edge of the Pacific Plate were the large islands that made up what geologists call the Insular Superterrane, including Vancouver Island and the Gulf Islands. Caught between these two large land masses, the little Pacific islands were crushed and crumpled, but here and there along what is now the Fraser Canyon, pieces of them survived. Marble Canyon is one of those pieces.

Its light-colored limestone cliffs tell part of its history; the limestone was formed perhaps two billion years before, on what were then shallow beaches. Here blue-green algae—which were in fact cyanobacteria—lived and died in unimaginable numbers. Their microscopic crystallized remains littered the sandy shores of the islands, creating reefs and, as they did in many other places, laid the foundation for a range of mountains, in this case the Pavilion Range.

Geologists believe these single-celled creatures did much more than create the first reefs and build the mountains of the future; through photosynthesis, cyanobacteria also helped to create the oxygenated atmosphere on which most of life on Earth depends.

Over time, the colonies of algae grew upward, layer upon layer, creating what geologists have called "biological felt". And in many places they formed rounded or mushroom-shaped colonies called stromatolites. Remarkably, stromatolites virtually identical to the fossilized remains are still being formed today

© Gary Bell / Oceanwide Images.com

Virtually identical to some of Earth's earliest creatures, these stromatolites thrive in the densely saline water of Australia's Shark Bay.

in a few locations, including the intensely saline water of Shark Bay, in Western Australia.

The stromatolites of Marble Canyon are particularly interesting because geologists believe they are a freshwater type that has only been studied in the past decade or so. Generally found in water that allowed intense calcification, particularly around hot springs

CYANOBACTERIA, ARCHAEA
Searching for our distant ancestors

Blue-green algae were the forerunner of plants and animals and for decades, scientists believed that they were the only inhabitants of the Earth's earliest environments. Now, powerful microscopes have revealed two other primeval forms of life. One, called Eubacteria, evolved into modern bacteria, while the other, Archaea, includes little-understood organisms that continue to live today in some of the world's most hostile environments: deep sea rift vents with temperatures over 100°C, or extremely acid or alkaline waters. Much has yet to be learned about these primeval forms of life. However, a consensus is emerging that these primitive microorganisms exchanged genes; if so, the concept of the single common ancestor for all life becomes rather more complicated than once believed.

or in heavily mineralized water, they have recently been found in a number of places around the globe, including the Bahamas, the Indian Ocean and Yellowstone National Park. Most are small, perhaps a couple of centimetres across, but the stromatolites found around Pavilion Lake were among the largest ever found.

Getting There: Marble Canyon Provincial Park, with its chain of lovely lakes, is 35 km northeast of Lillooet and 40 km northwest of Cache Creek on Hwy 99. The campground is open year-round and attracts rock and ice climbers, who call the often-overlooked canyon the "Cinderella of B.C. rock".

Peter St. John

The emerald waters of Crown Lake (below) and its neighbors, Pavilion and Turquoise Lakes—the latter with its lovely waterfall—as well as the canyon's limestone cliffs, draw visitors to the park year-round.

Keatley Creek

MOTORISTS PASSING OVER Keatley Creek on the Cariboo Highway north of Lillooet might be forgiven for failing to realize that they are but a short distance from the site of one of B.C.'s most remarkable ancient villages. Today, there is little to distinguish the creek from its neighbors to the north and south. Often dry during the late spring and summer months, it springs to life in the fall and winter, running out of the Clear Range Mountains to the east and across the sagebrush scrubland of the Fraser Canyon's upper east side, before plunging into the river.

What motorists can not see from the road is the fan-shaped glacial terrace east of the road. Stretching for almost a kilometre to the foot of Mount Martley, it straddles the creek bed. Sheltered from the wind, with abundant fresh water, for at least 1,500 years this was the winter home of one of B.C.'s largest, wealthiest and most socially complex communities.

Definitive dates from some of the largest earth lodges or pithouses indicate that the town was well established 2,600 years ago. And according to SFU archaeologist Brian Hayden, who spent more than two decades directing excavations at the site, there is evidence that a permanent seasonal settlement may have occupied the terrace even earlier.

The main town site, which at its height 1,200 years ago might have cradled a population of 1,500 or more, sits in a kettle depression in the terrace just north of the creek.

Outlying "suburbs" stretch east up onto a small upper terrace above the main site and south across the creek gully. All told, the town on Keatley Creek was almost a kilometre long and covered an area of about twelve hectares.

More remarkable than its size, however, was its social structure. Here, at its height, was a society that appears to have rivalled the complex, multi-levelled societies of Papua New Guinea. Hayden has termed this a Classic Lillooet social structure. These "Entrepreneur" communities, as he has dubbed them, have also been found in a number of other bountiful places on the Interior Plateau. They featured an elite group of powerful, wealthy families, a secondary social level of noble families, as well as shamans, warriors and traders, and common working folk, including fishers, hunters, carvers, basket makers, leather workers and perhaps slaves.

But why, travellers might ask, here at Keatley Creek? Today, the region seems dry and desolate, the soil too poor to sustain agriculture and too dry

Peter St. John

Though steep-sided and apparently inaccessible, the Fraser River in fact held the key to the town's success.

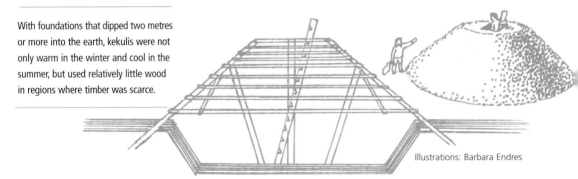

With foundations that dipped two metres or more into the earth, kekulis were not only warm in the winter and cool in the summer, but used relatively little wood in regions where timber was scarce.

Illustrations: Barbara Endres

for cattle grazing (though cattle were grazed on the ancient town site in the early nineteenth century). The reasons for this vibrant, remarkably long-lived community reflect the currency of early Lil'wat or Lillooet society, the geography of the region and, clearly, the commanding nature of generations of enterprising, charismatic and mutually supportive men and women.

The currency of this remarkable community was salmon, specifically spring and sockeye salmon, both large, rich, strong swimming species that once climbed the Fraser River in almost uncountable numbers every August and September in their quest to reproduce. The geography of the region includes the mighty Fraser, located below the townsite to the west, the Clear Range Mountains, which climb sharply just to the east and the neighboring ridges to the north and south, which offer protection from winter winds.

Though the banks of the Fraser are steep in many places, there are stretches of gravelly shorelines at the mouth of Keatley Creek. These would have been suitable for harvesting smaller, shore-hugging salmon. They would have been of little help, however, in catching the huge numbers of sockeye and spring salmon necessary not only to sustain a community the size of Keatley Creek, but—filleted and quickly dried in the hot canyon winds—to serve as valuable items for an extensive trade network.

Initially Hayden found it hard to believe that the inhabitants of Keatley Creek could have procured the fish they needed along this stretch of the river. Even at the Camelsfoot Constriction a kilometre upstream, where the Fraser narrows and rapids force the deep-swimming sockeye and spring or chinook salmon closer to the surface, the canyon walls are sheer and

high. Where could fishermen have positioned themselves to net the thousands of salmon needed?

There was only one way, Hayden decided after looking at many options; it involved scaffolding suspended precariously from the tops of the cliffs. The engineering needed to construct such ladders and platforms, not to mention the bravery needed to use them, is breathtaking even to contemplate. Yet thousands of salmon were caught this way every year by fishermen sweeping their nets through the water. At peak migration times, an experienced fisher might, with luck, catch between seven and eight sockeye and one or two of the less numerous spring with one sweep of the net. Given that the sockeye can weigh as much as five kilos apiece, it's likely that the fishermen would have had to take regular rest periods. Even so, working from similar structures on the Columbia River in more recent times, fishers were able to catch upwards of 300 salmon a day, each. One can only imagine the army of women and children needed to fillet and dry the fish during the peak migration.

While salmon constituted the majority of calories consumed, it was not the only food the Lillooet people ate. They also gathered and dried berries, roots and corms, and hunted deer in the mountains in the fall, which they ate both fresh and as dried jerky.

Hayden believes that the best fishing sites belonged, in perpetuity, to the elite families of Keatley Creek. Salmon were the currency that earned them their huge homes and allowed them their lavish lifestyles. Five of the 119 pithouses at Keatley Creek were significantly larger than the others, with floor areas of as much as 115 square metres—roughly equivalent to a house with six large rooms. Hayden estimated that about nine families, or an average of fifty people, lived in each of the great houses. The Lillooet termed this a *pel'u3em*, which Hayden terms a "corporate residential group".

Clustered around each great house were medium-sized and small pithouses, as though each aristocrat had his own congregation of lower-class families. In return for his protection and distribution of food and firewood, they performed all kinds of work, rather like tenants on the aristocratic estates of England once did.

To prove their wealth and status, great feasts were occasionally held

in the great houses, at which gifts—among them jewellery and exotic items from afar, tediously carved items of jade (see page 148), basketry, dog and goat wool blankets (see page 139) and beautifully tanned deer skins—were judiciously given to hold admirers and attract potential partners. The Lillooet were well known for preferring peace to war.

The children of the Classic Lillooet great houses were trained from infancy with as much care and precision as the offspring of Egyptian princes or English monarchs. Schooled for years, trained and hardened, they were often kept in relative seclusion for much of their adolescence. Boys were sometimes taught separately for up to ten years, while aristocratic girls were considered to be more valuable the better they were trained and the longer they were in seclusion, often up to four years prior to marriage.

This remarkable way of life, which had lasted for centuries, ended abruptly about 1,100 years ago when, one summer, the salmon failed to appear. Very likely the result of a landslide downstream, the fish upon which all depended simply did not come. One can only imagine the puzzlement, panic and ultimate pandemonium that resulted. But pragmatic to the end, the Lillooet people packed up their belongings— perhaps saving at least the best of the valuable roofing timbers—then torched all their homes and went away, likely to seek assistance with allies downstream or even farther afield.

The land they left stood empty for centuries, before people again began to populate the river shores. Some were Lillooet, from the west; others were Secwepemc from the north and east. But the terrace on which one of the great towns of early B.C. had proudly stood for so many generations was never again inhabited, except by cows. Today, more than a thousand years later, one can still almost hear the laughter of the children and smell the salmon cooking.

Those images may yet come alive, for there are plans for a Lillooet cultural centre at Keatley Creek; celebrated architect Douglas Cardinal has been hired to produce a plan.

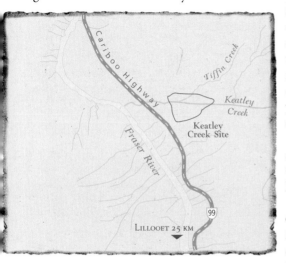

Getting There: Keatley Creek is located about 25 km upstream from Lillooet on Hwy 99; just north of Sallus Creek; the ancient town site is east of the highway on a terrace at the foot of Mount Martley.

145

The Stein Valley

THIS MAGNIFICENT VALLEY is awe-inspiring; one does not have to be Nlaka'pamux—"the people of the canyon"—to feel its power. But for today's members of the Lytton First Nation, who still live in small communities up and down the Fraser Canyon, as well as along the lower Thompson and Nicola Rivers, this is a holy place. It has been this way for thousands of years, before Moses went to the mountain, before Rome became the centre of Catholicism.

Set among spectacular peaks, including 2955-metre Skihist Mountain, the valley curves in an arc from just east of Lillooet Lake to the Fraser River. Its slopes are carpeted with ponderosa pine, cottonwood, birch, Douglas-fir, redcedar and spruce. At its heart is the Stein River, which tumbles in a series of falls and cascades from its headwaters at Tundra Lake in the high meadows at the western end of the park to pristine Stein Lake and then, gathering water and power from lakes and streams as it goes, winds almost sixty kilometres through the valley to pour into the Fraser River just north of Lytton.

But the Stl'yen, the "hidden place", as the people who have lived here for millennia call it, is more than breathtaking natural beauty. This is hallowed ground, a place, as the late elder and Nlaka'pamux spokeswoman Annie York said, "for dreaming". The evidence is everywhere. Along the beautiful valley, and up many of its tributaries are cliffs, boulders and caves where pictographs and petroglyphs—rock paintings and carvings—that attest to the valley's deep and lasting spiritual importance to its people. Most of the rock paintings are found on places that have been long recognized to possess high levels of spirit power. Many of these are on cliffs or boulders along the ancient trail beside the river, but some are in caves high on

Photos: Dennis Fast

With venerable trees cloaked in lichens and offerings in high niches, the valley gives visitors a glimpse of a world that no longer exists in far too many places.

the steep mountain slopes. Even today, instructed by shamans, young initiates come here to seek the wisdom of nature and the power of the past.

To protect this magnificent natural cathedral, Stein Valley Nlaka'pamux Heritage Park was created in 1995, setting aside

Stein Valley
Nlaka'pamux
Heritage Park

A natural cathedral, the Stein Valley elicits not only respect, but reverence.

To protect this cultural wealth, the Lytton First Nation and the provincial government have agreed to make the Stein Valley a living museum of cultural and natural history. But though the Nlaka' pamux people have opened the doors to their magnificent natural cathedral, they have done so in the expectation that visitors will preserve, maintain and encourage both its exquisite beauty, and its ancient and priceless cultural worth.

the entire 109,000-hectare watershed. The valley had been largely concealed from the world until the 1970s when, to keep it from being logged, the Nlaka'pamux and their western neighbors, the St'at'imc, who had long shared its hallowed ground, appealed to international environmentalists to help "save the Stein".

It was well worth saving. Not only is the valley beautiful, but the lower Stein is exceptional in terms of the density and diversity of its archaeological sites. Seventy-eight sites have been identified, a concentration believed to be greater than any other area in the Interior Plateau. In addition to the pictographs and petroglyphs, these include ancient trails, rock cairns and configurations—"legend rocks", as the elders call them—culturally modified trees (see page 288), trail markers, burial sites, hunting blinds and drive fences, birthing, puberty and battle sites, as well as camp and village sites.

Getting There: From the Trans-Canada at Lytton, take No. 12 north to the ferry across the Fraser River. Travel north on West Side Road for 4.5 km to the parking lot on the sandy bench lands. Two trails lead from the cairn in the parking lot; a short branching trail follows the east and west branches of the Stryen Creek, while the main Stein River Trail winds 58 km along the valley, climbing to Stein Lake. The trail is rugged and the valley is home to both black and grizzly bears. Aside from hand-operated cable cars over the river in two places, there are no support services along the trail. The suggested return time to complete the trail is about a week. Two shorter wilderness trails can be accessed from the Sea to Sky Highway (see page 199).

147

S P O T L I G H T

Tools of Jade

GEOLOGISTS call this dense, compact mineral nephrite, but most people know it as jade. One of the world's hardest and toughest materials, it is stronger than steel, has a remarkable ability to keep an edge and is virtually shatterproof. (Another material, also called "jade", is jadeite. A tough, heavy stone, it's actually harder than nephrite, but considerably more brittle. Jadeite is found only in one small area of B.C., as well as the Yukon and California, but has been mined for millennia in several countries, including China and Myanmar.)

For early occupants of the Fraser Canyon and the Interior Plateau, who used jade extensively to create chisels, adzes, axes and scrapers for cutting or shaping wood and treating hides, as well as for knives, hammerstones, pestles and even war picks, its strength had both advantages and disadvantages.

The advantages are obvious. Jade is durable and rarely breaks. It is also a gem stone, so that even when used for something utilitarian, its value outstrips its functional use. And when finely polished, even an axe or a chisel becomes a thing of beauty and worth.

The downside is that, for the very reasons it is so valued—strength, hardness and durability—nephrite is exceeding difficult to work. This was particularly so for cultures lacking the diamond drills and high-speed equipment that are used today to sculpt and fashion jade. Yet beginning almost 4,000 years ago on the Northwest Coast and along the lower Fraser River, and about 3,000 years ago on the Interior Plateau, jade began to be used to create woodworking tools—adzes, axes and chisels—as well as knives and hammerstones. These timelines were not purely coincidental. They reflect the stabilization of dependable salmon runs and mature redcedar forests, which led to semi-permanent or permanent

villages along the coastal estuaries, rivers and lakes. The houses in these villages were made either entirely of wood (as on the coast) or were wooden-walled and roofed earth lodges or

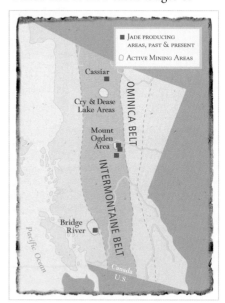

pithouses. To build them, people in many different cultures needed dependable, long-lasting tools. Nephrite filled that need.

Initially, because of the incredible hardness of the stone—hitting a jade rock with a steel hammer not only leaves no mark on the stone, but sends the hammer springing back with unexpected intensity—and the lack of both tools and time to work it, jade tools were quite crude, little more than modified nephrite pebbles. Over time, however, people adapted what archaeologists call a "ground-stone technology" to create more satisfactory tools. Using sandstone,

Above: Tools of jade, such as these from the Simon Fraser University and Sooke Museums, were greatly prized, for creating them took days or even weeks and they lasted for years.

granite or hard volcanic grinding stones in conjunction with abrasives like sand and water, they slowly sawed or drilled jade boulders into desired shapes. Nor was jade used strictly for tools. At a number of sites, the adzes, axe heads and knives that were found seemed disproportionately large or lacked any signs of wear. It seems they were created purely as status symbols, indications that the owner had the time, or the manpower, to manufacture such costly goods simply for ceremonial or display purposes, or perhaps to trade.

The remarkable length of time this hand sawing takes—modern experiments using traditional methods resulted in grooves that deepened at a rate of less than 1.5 millimetres per hour—and the number of jade tools found say several things about the

Understanding the traditional process for making groundstone tools, above, makes the end result (at right) even more remarkable.

cultures that created them. The first is that they had the resources to afford a relatively sedentary lifestyle. Almost invariably, this affluence depended on salmon.

Second, as Brian Hayden has proven in his study of Keatley Creek (see page 143), in some places it led to a stratified society in which certain families became very wealthy. And this led to an ability to "contract out" the wearisome task of grinding, either to poor members of the community or to slaves.

As the map indicates, nephrite has so far been found only near the outer edges of the Intermontane Belt, which bisects B.C. from south to north. Within this band, it can be found in outcroppings and boulder fields as well as pebbles and larger stones in the area's rivers. Because it is nearly impossible to remove from outcroppings without modern tools, early British Columbians found the jade they needed in riverine pebbles and boulders. Thanks to nearby deposits and the assistance of glaciers, these were most plentiful along the rivers and creeks just north of Lillooet.

Despite jade's narrow origins, the articles created from it travelled remarkable distances, another indication of its value. Artifacts of nephrite have been found from the northern coast to the eastern Kootenays, though the greatest concentrations are found in the Lillooet and Lytton areas, as well as up the Thompson Rivers to the Shuswap Lakes, for the Secwepemc people travelled and traded on a regular basis.

Fittingly, given its long history of use, jade is B.C.'s official mineral emblem and is mined in many places. The province produces about 100 tonnes of nephrite annually and supplies most of the world's requirements. Today, it is used mainly for ornaments, jewelry and sculptures; one of the most spectacular is the world's largest jade Buddha, which was carved from a thirteen-tonne boulder and now adorns a Bangkok monastery.

Though you have to be something of an expert to recognize them, jade pebbles and boulders can still be found on the gravel bars of the Yalakom, Bridge, Coquihalla and lower Fraser Rivers. These areas are open to public collection under the Fraser River Jade Reserve, which extends from the Fraser River bridge at Hope to the suspension bridge in Lillooet, and allows anyone to collect jade for his/her sole use and pleasure, below the mean high water mark, without the need for a Free Miner's Certificate.

149

Hell's Gate

BC Archives, A-03882, c. 1914

EVEN FOR THOSE who are familiar with it, Hell's Gate is breathtaking. A deep gorge that narrows to just thirty-four metres, its walls of rock soar 1000 metres above the water, which thunders through at a peak rate of 15000 cubic metres a second. For decades, scientists have been trying to decipher its long and fascinating history. But interpreting the geological history of the Fraser Canyon has been compared to reading a book in which most of the words are missing. Despite decades of investigation, many geological maps still show this region with question marks on it.

What geologists do know is that here, compressed, convoluted, faulted and fragmented are the remains of six ancient terranes—bits and pieces of the Earth's crust, each with its own geological history—that merged more than 100 million years ago. At that time—the mid-Cretaceous Period, or the Age of the Dinosaurs as some call it—these terranes were lying south of what is now British Columbia. In a complex series of plate movements, as the North American continent moved west to collide with the Insular Superterrane, which was riding on the edge of the Pacific Plate, it seems that fragments of the six ancient terranes were caught and squeezed between them.

This jumble of volcanic rock, gravels and clays was then compressed by time and pressure into limestone, sandstone and a rough cement-like rock called conglomerate. Fractured along a major break in the Earth's crust known as the Fraser Fault, this region of crushed rock was easily eroded, and prone to landslides.

Vancouver geologists John Clague and Bob Turner believe that prior to the beginning of our current ice age, about two million years ago, the Fraser was a small, coastal waterway with headwaters no farther north than the southern Chilcotin. On the other side of a height of land (which others believe may have been south of Lytton), a much larger, north-flowing river drained a huge part of northern B.C., perhaps through what are now tributaries of the Peace River.

When the southern Fraser captured its northern neighbor, very likely thanks to changes in the drainage patterns caused by advancing or retreating glaciers, the vastly increased flow acted like a drill on the rock of the Fraser Fault. From north of Lillooet through the lower Fraser Canyon, the water carved the canyon that visitors to Hell's Gate find so awe-inspiring.

Unfortunately, this region is also known for its landslides. As an indication of the frequency that rock avalanches occur, in the past four decades, approximately 3,500 landslides have occurred along the roads and rail lines in the two main transportation corridors in southwestern B.C.—the Fraser Canyon route between Vancouver and Kamloops and the Sea to Sky route between West Vancouver and Lillooet.

One of the most famous took place in 1914, when construction on the Canadian Northern Railway triggered a huge landslide at Hell's Gate. Thundering down the canyon walls, the rock slammed into the river, briefly damming it, spraying mud and debris far up the opposite wall and all but destroying the world's most productive salmon river. Previous blasting had largely blocked the river in 1913; the landslide made passage almost impossible. That summer, as the annual spawning run began, the salmon arrived at the barrier. Leaping and flailing, they launched themselves at it, desperate to reach their spawning streams. But only the strongest were able to surmount the piles of rock; the rest died in the millions. And upstream, the people, animals and birds that depended on the salmon run starved.

One indication of the magnitude of this disaster are the spawning numbers along the Horsefly River, before and after 1913. A beautiful tributary of the Quesnel River, the Horsefly originates just west of Wells Gray Provincial Park and has long been one of the great salmon spawning areas of the Fraser River watershed. In 1909, approximately four million sockeye were reported in the river's spawning grounds. That total dropped to almost zero in the years following the landslide and took decades to recover. In 1941, only 1,100 sockeye were recorded on the Horsefly River. In 1945, fish ladders were opened at Hell's Gate to assist the migration and the numbers began to climb. Still, the recovery continues. In 2005, the spawning count in the Horsefly River was estimated at 2.1 million fish.

Much earlier, the region had played a role in the dispersal of coho salmon. At the end of the last glaciation, melting ice created enormous proglacial lakes in southern B.C., which created connections between watersheds that are separate today. For a time, when the Fraser Canyon was blocked with ice near Hell's Gate, coho salmon and other species colonized the interior Fraser/Thompson River watershed through glacial lake connections in the Okanagan-Nicola areas and through connections between the Columbia and upper Fraser Rivers far to the west.

When the great glacial lakes disappeared, these connections were cut off. The result is an upper Fraser coho population that is genetically distinct from all other coho in Canada. Because of the length of the river and the number of its tributaries, this population makes up about twenty-five per cent of Canada's natural coho. Like their cousins downstream, interior Fraser coho migrate to the ocean and return to the streams of their youth to spawn, primarily within the traditional territories of the Secwepemc, Nlaka'pamux and Okanagan people. Some also spawn in the Lillooet-Bridge River areas and in the Chilcotin river system.

Getting There: For those who take the old Trans-Canada Highway route along the Fraser Canyon (rather than the newer and faster Coquihalla Highway), Hell's Gate is hard to miss. Located about 12 km south of Boston Bar, and well signed, it features an airtram that takes visitors from the upper parking lot down to the rushing water where viewing decks, gift shops and restaurants can be found. The airtram runs between mid-April and mid-October. A fee is charged.

The Salmon

THINK OF THE FRASER RIVER, and salmon almost invariably come to mind. For more than 5,000 years, this river has been among the world's foremost producers of these remarkable fish and entire cultures have been built around the coming of the salmon. The enormous, dependable runs allowed the evolution of sophisticated societies and the creation of art that today symbolizes and distinguishes British Columbia not only across Canada, but internationally. It is a sad irony, therefore, that both the cultures who produced these distinctive art forms and many of the species on which they depended are today threatened.

Though paleontologists and archaeologists believe the enormous spawning runs—salmon returning in numbers large enough to reliably support societies for months at a time—go back five or even six millennia, salmon as members of a species are far older than the river itself. The ancestors of today's salmon date back to the Early Cenozoic, between fifty and sixty million years ago. Fossils of a very early relative of the family *Salmonidae*, which includes salmon, trout, char, grayling and whitefish, have been found in fossil beds along the Driftwood

Creek near Smithers (see page 112), where until recently huge numbers of spawning salmon could still be found. In short, the salmon are almost as old as parts of the province of British Columbia. Much younger fossils, perhaps 18,000 to 21,000 years old, have also been found on the south shore Lake Kamloops, raising interesting questions about the extent and timing of the last glaciation (page 108).

Today, there are six distinct types of anadromous Pacific salmon, ranging in size from the huge chinook or spring salmon, which weighs up to fifty kilograms and lives between five and seven years, to the little pink, which weighs 2.3 kilos or less and lives only two years. Perhaps best known is the sockeye, with its epic struggles upriver to spawning grounds hundreds of kilometres inland. There is also a freshwater form of this species, known as kokanee salmon (page 121).

Though all Pacific salmon begin and end their lives on the gravelly beds of cold mountain streams, they are also creatures of the ocean. And despite years of effort, scientists know relatively little about their lives at sea. Ocean tagging has produced some remarkable insights, however; a sockeye tagged at 177E West, a longitudinal parallel that runs through eastern Siberia, south through the Pacific east of Kamchatka and Japan, was found later in the same year in the Nass River, a journey of at least 2,000 kilometres. Known for its prodigious freshwater travel, it seems the sockeye also clocks the greatest distances in the four or five years it spends at sea.

It seems that some salmon may have survived the last glaciation in

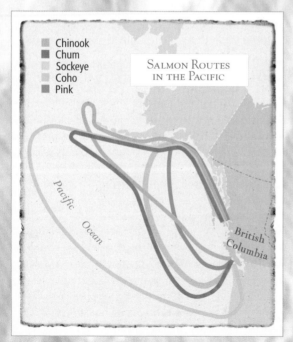

SALMON ROUTES
IN THE PACIFIC

Chinook
Chum
Sockeye
Coho
Pink

Pacific
Ocean

British
Columbia

B.C., for salmon bones dating to 16,000 BP have been found at Port Eliza, on Vancouver Island's west coast and from a site in Haida Gwaii nearly 13,000 years ago.

After the disappearance of the ice about 11,000 years ago, a long period of global warming known as the Hypsithermal (see page 20) existed for more than 4,000 years. Following the cold water, the salmon stocks shifted north. It was not until about 6,000 BP, according to painstaking work by archaeologists, that the stocks had recovered enough to constitute a large part of the diets of early British Columbians. And it was even longer before salmon could sustain large populations. Early riverine communities likely caught salmon when they were available, but for millennia they were not a staple, but rather an alternative to a diet that was mainly meat or marine protein, plants and fruit.

The story of B.C. salmon, and particularly the tale of the sockeye, is nothing short of remarkable. Hatched in early spring from gravel beds in clean, oxygen-rich tributary streams, small rivers or lakeside beaches, each tiny fry has a pendulous yolk sac that sustains it for the first two or three weeks. Then it's on its own. Depending on the species, salmon young, called smolts, spend up to three years in freshwater lakes and rivers before the travelling to the ocean in May and June. Once there, again depending on the species, they migrate hundreds or thousands of kilometres out into the northern Pacific. Scientists are not yet sure where each population goes. Increasingly however, thanks to genetic testing, scale growth rings, parasites and other identifying markers, they do know where each salmon

caught in the open ocean originates.

Sockeye—the word may come from the Coast Salish *sukkai*—are the second most abundant species in British Columbia waters (after the smaller pinks) and the most economically valuable. They head for the ocean in the spring of their second year; there the young adults—silver in colour, with distinctive scales, dark tails and black speckled backs—move northward along the coast. Unlike other Pacific salmon but like some whales, sockeye are filter feeders, feeding on plankton strained from the water.

For the next two years, they cruise the ocean, apparently travelling enormous distances. Then, at four or occasionally five years of age, the urge to spawn overcomes each sockeye and it heads for the rivers and streams in which it was born. Scientists are still trying to understand the spawning impulse, for it has obvious implications, both for the fish and the populations who have long depended on it. Investigations conducted during the past fifty years have shown that salmon are able to use the sun's position and Earth's magnetic field, among other positioning systems, to orient themselves at sea and, once in fresh water, each fish follows cold water and can "smell" the distinctive chemical odor of its home stream. As they head upriver in the late summer and early fall, the bodies of sockeye turn bright red, their heads green and the males develop large hooked jaws.

Once the urge to spawn takes over, each fish will either overcome every obstacle to reach its final destination, or die trying. Each year's run is different, with dominant runs occurring every fourth year. Before the coming of Europeans, so many fish crowded the smaller tributaries during these bountiful years that some were actually squeezed out of the water onto the banks. And every year, the narrows and rapids of the Fraser and its hundreds of tributaries were crowded, not only with people, but with bears, eagles, raccoons and many other species whose lives had come to depend on these remarkable fish.

THE
FRASER
VALLEY

Peter St. John

The Lower Marsh in Minnekhada Regional Park reflects its name, which means "beside still waters" (though rather incongruously in a Siouan language). The park is known for its birdlife; visitors often see cranes, bald eagles and hawks.

Like a skeleton's smile, the remnants of a old weir puncture the Fraser River silt at the mouth of the Salmon River in Derby Reach Regional Park. Named for the stretch of the river between Fort Langley and Barnston Island, the park is known for Edgewater Bar, still a favorite site for anglers.

The Fraser Valley

Broad and fertile, edged with wetlands,

bordered by mountain streams and bisected
by one of the most productive rivers in North America,
the lower Fraser Valley is testament to the power of ice and water.

As IT UNDOUBTEDLY has before, glacial ice filled the valley about 18,000 years ago, forcing the many animals that called it home to move south and west. Among them were enormous Columbian mammoths, which stood nearly four metres high at the shoulder and had tusks almost as long. Seeking the forage they needed, they may even have crossed the Strait of Georgia, which—thanks to sea levels that were almost 100 metres lower—was much narrower than it is today. Washington's Orcas Island and San Juan Islands could have provided stepping stones to the Saanich Peninsula (see page 231), which remained unglaciated for at least another thousand years.

For the next six thousand years, the Fraser Valley was buried deep in ice. When the great freeze finally ended about 12,000 years ago, the ice began its long retreat, depositing huge boulders along the edge of the valley and leaving deep deposits of sand, gravel and silt along its bottom.

One of the boulders took on a special significance almost immediately. By 9,000 BP, people were camping in its shadow. Ultimately, it would be given a name—Xá:ytem—(see page 164) and be recognized as the centre of an ancient community. Today, it marks a heritage site that boasts superb reconstructions of the pithouses that once sat above the river's edge, as well as a vibrant interpretive centre that offers workshops and activities throughout the year.

Elsewhere, the glacial sediments formed isolated mounds, called kames, and small hills, known as eskers. These elevated points included one that would be home to early British Columbians for more than 8,000 years, and eventually the site of Fort Langley (see page 166), which is today a National Historic Site.

Even when the ice was gone from the valley, meltwater continued to carry enormous loads of sediments from the mountains and valley inland to the sea, just as the Fraser River does today. By 10,000 BP, this sand and silt had begun to pinch the mouth of what geologists call "Pitt Fjord" (see page 168), a long arm of the sea that stretched deep into the mountains on the north side of the valley. Rapidly building wetlands and marshes, it transformed the fjord, creating Pitt River and long, narrow Pitt Lake, North America's largest freshwater tidal lake. Here, thanks to the

The Fraser rolls past Derby Reach, (right), briefly the site of the first incarnation of the Hudson's Bay Company's Fort Langley.

Fiddleheads (centre) are prized for their fresh taste of spring, just as they were thousands of years ago.

Photos this page: Allan Taylor

combined efforts of aboriginal and non-aboriginal residents, the wetlands and uplands of Pinecone Burke Provincial Park, as well as the rare peat lands of Blaney Bog have been set aside as havens for wildlife.

Farther east, scars on Mount Cheam (see page 161) mark what geologists believe was the worst rock avalanche in British Columbia history, a slide that may have been triggered by water pressure 5,000 years ago. Nearby, as if to serve as counterpoint to that act of earthly violence, is delicate Bridal Veil Falls (page 163). And through it all, the wide Fraser River flows, continuing its endless job of building its delta into the strait.

Kanaka Creek, east of Maple Ridge, winds toward the Fraser. The sinuous stream was long a spawning site for salmon; today its shores boast hiking and riding trails.

The Fraser River

Illustrated London News / National Archives of Canada / C-121041

MENTION THE FRASER RIVER, and people may think of Vancouver, or Hell's Gate in the Fraser Canyon, or perhaps Prince George near the river's most northerly extent. The Fraser is all of these and much more. Rising on the Pacific slopes of Fraser Pass in Mount Robson Provincial Park, it flows 1375 kilometres to the coast, making it Canada's fifth-longest river. Its watershed is greater than the area of Great Britain, and for millennia it has incubated the most productive salmon fishery in the world; as a result it has been one of the focal points of life in British Columbia for several thousand years. It is the Stó:lo, as the people who were named for the river call it— not just any river, but "*The* River".

Yet geologists believe that the Fraser has not always been such a lengthy, significant, productive river. After studying it for decades, they have come to the conclusion that the river we know today is a youngster, geologically speaking, and that it grew to its present size through a process known as stream piracy.

Prior to Earth's most recent ice age, which began about two million years ago (see page 16), the Fraser seems to have been a small coastal river with its head-waters near today's Lillooet. Central and eastern B.C. were drained by a much larger, north- and east-flow-ing river that may have been the ancestral Peace.

As geologists John Clague and Bob Turner write in *Vancouver, City on the Edge*: "Evidence of this ancient river system can be seen in anomalies in the present pattern of drainage. Many tributaries … flow northward, rather than southward as one might expect if the Fraser had always drained south …" Other evidence can be seen in the Fraser Canyon, which is deeper than it is wide in places, unusual for a major river, and in the sand and gravel bluffs along the river south of Prince George, which appear to have been "deposited by a river that flowed north", according to Clague and Turner.

The Fraser was likely aided in its piratical behaviour by the bulldozing effect of glaciers during one or more of the glaciations that have invaded the Northern Hemisphere over the past two million years. The enormous sheets of ice had the capacity to alter or even remove the drainage divides that separated the ancestral Fraser River from its much larger north-flowing neighbour, but the expansionist Fraser may also have been helped by other factors, such as vol-canic eruptions, mountain building and natural erosion. Whether the river expropriated its additional

Allan Taylor

One has to search for them, but there are still places where one can imagine the Fraser the way it was when the illustration (above) was painted a century ago, and even places that are much as they were ten centuries ago.

territory over a lengthy period of time, or captured it all quite rapidly, is not yet clear.

What is clear is that the Fraser is unequalled by any river anywhere in its innate capacity to provide rearing and spawning habitat for all six species of salmon indigenous to the Pacific—not to mention twenty-nine other species of freshwater fish and eighty-seven more found in the estuary. No wonder the river has been a crucial part of British Columbia's long human history.

Unfortunately, the past century or so of human history has been largely one of devastation. The destruction of spawning streams and crucial wetlands; chemical and biological pollution; uncontrolled logging, mining and pulp-and-paper making; rampant development, particularly on the river's flood plains, and excessive overfishing have together brought salmon stocks to the edge of disaster and put enormous pressure on many other species of animals, from the great blue heron to the white sturgeon (see page 74).

If there is a bright side to this sad recent history, it is in the efforts of dozens of organizations and agencies focused on trying, in one way or another, to repair the damage that's been done to the Fraser— and by extension to its ancient creatures—over the past 150 years. From volunteer groups like that in Salmon Arm that has replanted fifteen kilometres of the banks of the Salmon River, a small but once bountiful tributary of the Fraser, and another in Kamloops, which has breathed new life into a dying marshland, to native fishing officers who monitor the salmon as they lunge upstream in their age-old urge to spawn; from school groups that pick up the litter scattered about the riverbanks to governments that institute stricter environmental guidelines for industries that pollute the river, from university experts that study the problems to centres focused on public education, thousands of people have begun to take an interest in saving British Columbia's most important natural lifeline. One can only hope they're not too late.

Allan Taylor

With two kilometres of trails and raised walkways, the Cheam Lake Wetlands provide an opportunity to see many kinds of birds; 173 species have been sighted here.

The Mount Cheam Slide

TRAVELLING FROM HOPE to Vancouver on the Trans-Canada Highway, it's easy to miss the remains of the avalanche of rock that came down from Mt. Cheam about 5,000 years ago. The slide area seems less forbidding than other, more recent landslides, such as Alberta's 1903 Frank Slide. There, the evidence is stark and haunting, "a boneyard sealed by the dead weight of numberless slabs of limestone the size of houses", as Doug Whiteway described it in *In Search of Ancient Alberta*. Or the 1965 Hope Slide, which roared across the Crowsnest Highway in January of that year, killing four motorists and closing the highway for twenty-one days.

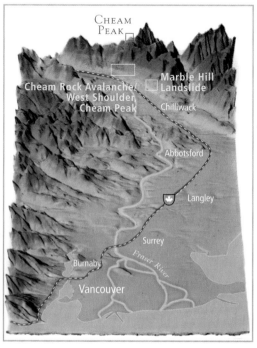

Yet the Mt. Cheam Slide was much larger and—according to the oral history and the legends of the Stó:lo Nation—far more disastrous than any in recent history. In fact, as John Orwin, John Clague and Robert Gerath have recently written, the "Cheam rock avalanche … is the largest known catastrophic landslide in western Canada". It released 175 million cubic metres of volcanic rock, limestone, conglomerate, shale and clay, which thundered down both the east and west slopes of Mt. Cheam, possibly simultaneously.

The slide on the west side of the mountain was particularly devastating; the debris roared across the alluvial terrace of the Fraser River, a site that had long been favored for fishing camps and villages, and into the Fraser River channel. According to the oral history of the Stó:lo Nation, the avalanche of rock and mud completely buried a village. And Keith Carlson, who edited *A Stó:lo Coast Salish Historical Atlas*, says that as a result, even today the Stó:lo people consider the avalanche site *xaxa* or "taboo".

Located in a region of volcanic and sedimentary rock that flanks the harder and more stable granite and gneiss of the northern Cascade Mountains, Mt. Cheam is composed of rock that dates to between 400 and 200 million years. This complex assortment of types and ages of volcanic rock, limestone and loosely cemented gravel was folded, faulted and jointed when the Coast Belt rammed into the rest of British Columbia about 140 million years ago. The result is mountains in the region that are inherently unstable.

A corresponding story is told about the Frank Slide in the western Crowsnest Pass, which buried the mining community of Frank. Though that slide is believed to have been triggered by mining tunnels at the base of the mountain, Turtle Mountain was also unstable, something that the people of the region had long known. Travelling back and forth through the pass for millennia, the Ktunaxa people called it "The Mountain That Walks" and never camped at its base.

Mt. Cheam was apparently made further untenable by glacial action between 18,000 and 11,000 years ago. The glaciers undercut the slopes, leading to grades that were too steep to sustain the weight of rock above, and also created drainage problems, which ultimately led to water pressure on weakened and fractured faults. It seems that Mt. Cheam suffered from both problems, for the east and west slides are rather different in character and may have occurred at different times. Geologists believe the enormous slides could only have been triggered by two things: an earthquake or water pressure from rainfall, snow melt or a pond near the source of the slide, on a fault or fracture. For the west slide, in particular, they opt for the latter. It seems that the source zone for the western slide was in the marshy basin of Bridal Veil Creek and it was the initial movement of waterlogged earth that triggered a much larger failure lower on the slope.

Mt. Cheam's eastern slide, also known as the Popkum Slide (for the Salish word for the "puff balls" that grow in the area) may have been triggered by the action on the western slope, or may have happened independently. Whatever the timing, when the unstable rock on the upper slopes began to slide, it roared across a lake or marsh at the base of the mountain, spraying muddy debris along its leading edges. This is much like the Hope Slide, which had dry rubble on the upper sections and muddy debris below.

Whatever their makeup, it is very unlikely that anyone in the path of either slide would have had any chance to escape, for rock avalanches of this type travel at unbelievable speeds. Again, quoting Doug Whiteway, "It took only 100 seconds … for the Crowsnest Pass at

the town of Frank to turn from green valley to sepulchre." The same was likely true of the slopes of Mt. Cheam.

And what of the aquatic inhabitants of the Fraser River itself? Looking now at the water's edge and the islands that straddle its width, it seems clear that the river has cut through the toe of the slide over the past 5,000 years. And it may well be that, like the Hell's Gate avalanche of 1914, which was triggered by blasting during the building of the CNR (see page 150), the Mt. Cheam slide partly or even mostly blocked the Fraser for a time, which would have had a disastrous affect on the annual salmon run and, by extension, on the people up-stream who had come to depend on the river's bounty for their livelihood.

Getting There: Mt. Cheam still towers 2107 metres above the Fraser Valley just south of the Trans-Canada Hwy about 30 km east of Hope. In addition to Bridal Falls Park described on the following page, Cheam Lake Wetlands Park to the west offers picnic areas and marsh, lake and uplands to explore. Cheam Mountain Trail and Cheam Peak Trail allow hikers to scale Cheam Ridge on the west side.

John Harvey

Five thousand years later, the scars of the avalanche can still be seen in this view of Mt. Cheam.

Photos: Allan Taylor

Bridal Veil Falls

THOUGH ITS ENGLISH name is not particularly imaginative—North America boasts more than a half-dozen waterfalls known by the same name—this lovely, lacy veil of water is worth seeing, particularly between March and June, when the flow rate is at its peak.

This is Canada's sixth-highest waterfall, cascading 122 metres (or nearly 400 feet) over a smooth volcanic rock face into Bridal Creek. The surrounding forest of western redcedar, western hemlock, maple and alder climbs low mountains that were rounded and smoothed by the last glaciation.

The region had an earlier name, inspired by the puffballs that still grow in the forest's sparse under storey. These fungi, which thrive in open woodlands and are edible when young, gave their name to the nearby town of Popkum, as well as one of two devastating landslides that thundered down from nearby Mt. Cheam.

Tiny Bridal Veil Falls Provincial Park is located sixteen kilometres east of Chilliwack. A day-use area that offers parking, picnic tables and hiking trails, it is easily found by following the signs from the Trans-Canada Highway. Bridal Creek flows through the park and into Cheam Lake Wetlands Park before joining the Fraser River.

It's unlikely that any of the other falls by the same name deserve it more than B.C.'s entry. Not only is the spray from the falls beautifully reminiscent of a bride's delicately gauzy veil, it is useful as well, allowing lichen, usually found in more humid conditions, to thrive in the park.

Xá:ytem

Brought here by the glaciers, long treasured by the Xat'suq people, Xá:ytem today presides over a magnificent reconstruction of the province's oldest known dwelling (at right, bottom image). Not far from the great rock is a smaller stone, called the Turtle Head (right inset), with what appear to be raised eyes.

JUST EAST OF MISSION, a massive boulder rises from the earth on the north side of the Fraser Valley. Geologists call this a glacial erratic, brought here by the enormous sheets of ice that once covered the valley and left behind when they began to retreat about 11,500 years ago.

The people who have long lived along the Fraser River have another name for it however; they call the great rock Xá:ytem (pronounced Hay-tum)—meaning "sudden transformation". They tell a story of three *si:yams*, or "respected leaders", who were instantly transformed for defying the wishes of the Creator.

More recent arrivals to the valley call this Hatzic Rock, a translation of the name of the Xat'suq, "people of the bulrushes", who have lived here since the Fraser River lapped at the south edge of the site about 9,000 years ago.

By whatever name, this is a place with a palpably powerful history. And today, thanks to the vigilance of archaeologist Gordon Mohs, the commitment of the Stó:lo Tribal Council, the cooperation of a Calgary developer, the assistance of the provincial government and the tireless efforts of director Linnea Battel, Xá:ytem has been resurrected, providing a remarkable window on the Fraser Valley's distant past.

It was Mohs who rediscovered the site. A firm believer in the archaeological antiquity of the Hatzic Rock area, he was particularly concerned when he learned of plans to split the enormous boulder to make way for a new subdivision. As the bulldozers began work in the fall of 1990, he took to strolling the area. On Thanksgiving Day, he spotted several stone tools in the upturned soil. When a brief survey revealed dozens of artifacts that had been exposed by

the bulldozer, the developer agreed to stop further work until an extensive archaeological survey could be completed.

But few were prepared for what the 1991 excavation would reveal. The site was found to cover at least 41,000 square metres and to contain remains of two pithouses that were first built between 5,650 and 5,300 years ago, making them B.C.'s oldest known dwellings. Constructed just after the rise of the Sumerians around the Persian Gulf, and before Minoan culture appeared on Crete, further excavations made it clear that the dwellings at Xá:ytem were regularly occupied and rebuilt over the next 300 years.

Excavating one of the houses, the archaeologists found it to be large—about eight metres by nine—originally constructed into the side of a gravel slope, and supported by magnificent cedar posts. The floor was of packed sand that had been brought in from somewhere off site and a large hearth, almost four metres across, was uncovered near the centre of the dwelling. Beyond the house and running parallel to the north end was what appeared to be a drainage trench.

Among the more than 16,000 artifacts unearthed during the

Peter St. John

164

XÁ:YTEM

excavation, were pebble tools for woodworking, stemmed projectile points and several artifacts of obsidian. Originating in Oregon, hundreds of kilometres south, the obsidian proves that long distance trade routes were in use more than 5,000 years ago, during what archaeologists call the Charles culture period. To add to the site's significance, radiocarbon dating on a piece of charcoal showed that earlier cultures had also used the site 9,000 years ago. At the time, land now more than seventy kilometres from the Fraser Delta would have been almost ocean-front property.

The antiquity, richness and extent of the site convinced the provincial government to bring it under the umbrella of B.C. Heritage. Since then, Battel's fundraising campaigns and a series of grants have underwritten the reconstruction of two magnificent pithouses, and the creation of an interpretive centre. The centre serves as a focal point for dozens of activities, from school tours to craft workshops—in Salish weaving, basketry and drum-making, among other things—as well as special events throughout the year.

Getting There: Located at 35087 Lougheed Highway (No. 7), just east of Mission, Xá:ytem can be accessed from south of the Fraser River via Hwy 11, the Mission-Abbotsford highway, and Hwy 9, from Rosedale to Agassiz. Both highways connect to Hwy 7. The site, which boasts what has been called "the best native gift gallery in the Fraser Valley", is open Monday to Saturday, from 9 a.m. to 4:30 p.m. year round, with extended summer hours.

Peter St. John

Allan Taylor

Fort Langley

THOUGH THE FOCUS of Fort Langley National Historic Site is clearly on the fort's role in British Columbia's fur trade and settlement history, there is much more to this site, for its geological and archaeological history goes back not just hundreds, but thousands of years.

Climbing the gentle slope to the main gate, it becomes clear that Fort Langley sits on a knoll. This mound of sandy soil may be a glacial kame, deposited by retreating glaciers between 12,000 and 11,000 years ago. Kames —isolated deposits of silt and sand—often formed in holes at the edge of the retreating ice, which later melted in place around them. Just south of Fort Langley is a series of small hillocks called Forest Knolls, which may either be additional kames, or perhaps the extension of a small glacial esker. Alternatively, these knolls may be the remains of an ancient delta that was later incised and eroded by meltwater.

Certainly there was water—and lots of it—in the wake of the melting ice. According to Vancouver geologists John Clague and Bob Turner, water spilled off the decaying edge of an enormous tongue of ice

Today, Fort Langley celebrates the fur trade, and the post's important role in the creation of the province of British Columbia.

that filled the Fraser Valley east of Abbotsford. The ice and water carried sand and gravel, which collected in deep deposits on the uplands between Langley and Abbotsford, as well as north of the Fraser River between Pitt Meadows and Mission.

As the glaciers retreated, melt-water looped around the south side of Fort Langley, almost encircling the little hill, before pouring into a long arm of the Strait of Georgia that once stretched far up the Fraser Valley and curved north into what geologists call Pitt Fjord (see page 168).

The Fraser River eventually cut across the north side of the looping meander around Fort Langley, turning the old river channel into an oxbow that slowly filled with sediment. While this is not easy to see from the ground, it is very clear from the air.

Not surprisingly, perhaps, given its riverside location and its elevation above the surrounding area, Fort Langley has been home to many

cultures over the past 9,000 years. And early British Columbians undoubtedly took up residence here for many of the same reasons that the site later suited the employees of the Hudson's Bay Company. The river provided both sustenance and a highway, while the site's elevation allowed both protection and an excellent perspective.

Archaeologists have found traces of at least five cultures—from early hunter-gatherers to residents of a pithouse village, much like the one upstream at Xá:ytem or the dozens that have been excavated in the B.C. Interior. The earliest occupation, obtained from a piece of charcoal in a hearth almost a metre below the surface, dates back more than 8,400 BP, making this the second-oldest site in the Fraser Valley yet uncovered.

The excavations also provided evidence that the fur traders who worked here between 1839 and 1858 were employed very much in the spirit of the place. Among the thousands of artifacts recovered were several of obsidian; analysis indicated it came from quarries hundreds of kilometres away, in southern Oregon. In fact, located between the plateau cultures to the east and the maritime peoples to the west, it's likely that throughout its long history, the inhabitants of Fort Langley were not only accomplished traders, but at home in several cultures.

Getting There: Fort Langley is just north of the Trans-Canada Hwy between Abbotsford and Coquitlam. Follow the highway to the 232nd Street North exit. Follow Parks Canada's "beaver" signs along 232nd Street to Glover Road. Turn right and continue into the village of Fort Langley. At Mavis Avenue, just before the railway tracks, turn right. Fort Langley National Historic Site is located at the end of the street. This incarnation of Fort Langley is the second; the first, the original capital of British Columbia, which was occupied between 1827 and 1839, was located 1.6 km downstream at the mouth of the Salmon River. Today the Fort to Fort Trail, a section of the Trans-Canada Trail, links the two sites.

Photos on these two pages: Allan Taylor

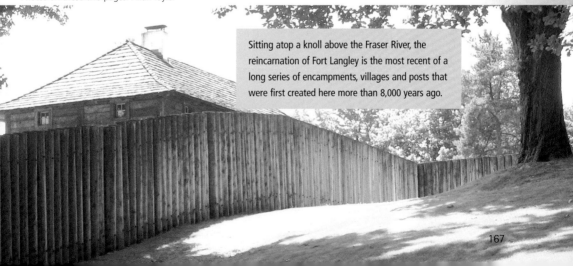

Sitting atop a knoll above the Fraser River, the reincarnation of Fort Langley is the most recent of a long series of encampments, villages and posts that were first created here more than 8,000 years ago.

Pitt Fjord

THIS AREA IS MANY THINGS to many people and its various entities have many names and a fascinating history that is still being written. Residents call its riverside communities, Pitt Meadows and Maple Ridge, the Fraser Valley's "best-kept secrets". The passionate, ecologically minded members of the Pitt Polder Preservation Society—*polder* being Dutch for "land reclaimed from the sea"—call Blaney Bog, the region's significant wetland, "a natural treasure". And geologists refer to the long, narrowing mountain valley, which stretches north and west from the Fraser Lowlands toward Garibaldi Provincial Park, as Pitt Fjord. In fact, this area is all these things—and more.

For almost a millennium at the end of the last glaciation, this was an ocean fjord that stretched far inland from the Strait of Georgia. In the earliest

Salmon weirs were built along many of the Fraser's tributaries.

centuries after the ice began its retreat, the sea covered all of what is now the community of Pitt Meadows, as well as the lowlands north of it, and lapped at the shores of the island of Maple Ridge. But the meltwater pouring off the rapidly disappearing Cordilleran ice sheet was filled with rocks, sand, mud and silt. As the rushing water met the sea, it slowed abruptly, dropping its huge load of sediment and rapidly building a huge marshy delta at what is now the confluence of the Pitt and Fraser Rivers. Even today, according to *Geology of British Columbia: A Journey Through Time*, the Fraser annually carries an estimated twenty million tonnes of sand and silt from the mountains and plateaus of eastern and interior B.C. and dumps it into the Strait of Georgia.

By 10,000 BP, the growing delta had pinched the mouth of Pitt Fjord, interfering with the tidal flow and creating ancestral Pitt Lake. For the next 5,000 years, the delta expanded, gradually shrinking the size of the lake and creating the Pitt River. Despite this, the effect of the ocean could still be felt; in fact, even today the Pitt River is an intertidal river and long, narrow Pitt Lake is North America's largest freshwater tidal lake.

Surrounded by glacier-topped mountains, fed by aquamarine streams, protected by huge marshes, with rich lowlands and easy access to the Fraser, this has long been one of B.C.'s most bountiful regions. Sturgeon, salmon and eulachon flourished in the river, berries grew

Amanda Dow

Human figure bowls are masterpieces in pecked stone.

along the shores and wapato—a semi-aquatic perennial that produces edible, starchy golf-ball sized tubers—grew in the marshes. The wetlands attracted moose, as well as large numbers of ducks, geese and other birds that are rare or endangered today, including greater sandhill cranes, great blue herons, tundra swans and peregrine falcons.

With such abundance, it's easy to see why the Katzie people and their ancestors have long lived around the lake. Even their name is linked to the region. Katzie comes from *q'eyts'i*, a Salish word describing the action of a human foot pressing down on moss.

Yet, despite millennia spent living and hunting in the region's wetlands, the Katzie left traces of their passing almost as ephemeral as their footprints in the moss. Occasionally, Katzie hunting parties would venture into the mountains in search of mountain goats. Pictographs on the sheer rock face above the southwest side of the lake detail such hunting trips and make it clear that both the hunters and the artists must have been as surefooted as their prey.

Even after European settlers moved into the Lower Mainland, the Pitt Meadows area, which was particularly flood-prone, was slow to be settled. But eventually riverside dikes made farming possible and by the late twentieth century, it was clear to both aboriginal and non-aboriginal

residents of the region that significant changes were taking place. To save both the Katzie culture and the land they had tended for so long, in 1995, 38,000 hectares of marsh and alpine territory along the west shore of Pitt Lake were protected within the boundaries of Pinecone Burke Provincial Park, to be managed jointly by the Katzie people and the provincial government. And in 2000, following a long battle against developers and the province that went all the way to Canada's Supreme Court, Blaney Bog—described by David Bellamy, a world authority on peatlands, as a completely intact "baby bog", with "all its fingers and toes", or elemental parts—was set aside for its rare birds, fish and plants.

Getting There: Pitt Meadows and Maple Ridge are off Hwy 7 northeast of the Fraser and Pitt River confluence. To reach Pitt Lake and Pinecone Burke Park, with its alpine creeks, back country trails and hike-in campsites, cross the Pitt River and take Coast Meridian road to Harper, which winds into the park's southern entrance. Or go east off Meridian onto Victoria Drive; follow it and Quarry Road north.

To paddle in the marshes at the south end of the lake, turn off Hwy 7 at the Harris Road stoplights, where a large sign points to Pitt Lake. Harris Road meanders north, then east. Turn north again at Neeves Road, following it across a narrow bridge over the Alouette River, where the road, now called Rennie Road, heads north through Pitt Polder along the Pitt River. Grant Narrows Park lies at the end of the road. Featuring a boat launch, it serves as a gateway to several wilderness marsh areas. Because strong winds spring up, paddling on the lake is not recommended. On request, the Katzie run powerboat and war canoe tours into the area. And visitors are welcome in the third week of August for Slehal tournaments featuring traditional gambling games that once had beautiful goat-hair blankets as stakes.

Blaney Bog is located north of 224 Street, which runs north of Hwy 7. There is no public access but walkers can view wildlife from a nearby dyke.

The Pitt River sparkles over the rocks.

169

THE
LOWER
MAINLAND

There are places, such as this stretch of Crescent Beach, where, looking out over the water, it's still easy to imagine life on the Lower Mainland as it once was.

Peter St. John

171

The view from Iona Island—newly-minted land, geologically speaking—across to Musqueam land and the site of the venerable Great Fraser Midden allows a fascinating juxtaposition of new and old.

The Fraser Delta

The view west from the heights

of Surrey today is over the apparently

endless lowlands of Delta, Ladner and Richmond.

IT'S HARD TO IMAGINE that all this land, home to a half-million people, thousands of businesses, hundreds of farms and both Vancouver International Airport and the BC Ferry Terminal, is, in geological terms, brand new. For the early inhabitants of the slopes of the Surrey highlands, the view west was a sea view, with the uplands of what are now Vancouver and Burnaby to the north and, off in the distance, Roberts Island, at the time one of the Gulf Islands.

Today Roberts Island, the Tsawwassen Upland, has been joined to the mainland by the enormous Fraser River delta, an almost feature-less plain that stretches west from Surrey far into the Strait of Georgia.

In their book on the city's geology, *Vancouver, City on the Edge*, co-authors John Clague and Bob Turner call the Fraser delta "Mother Nature's dump" —and from a geologist's perspective they are right. But paleontologists and archaeologists might argue that this is Mother Nature's priceless archive, for here, buried beneath layers of sand, silt and peat is a remarkable record of the past.

The layers of sediment have allowed geologists to reconstruct the advances and retreats of the dithering Cordilleran ice sheet between 21,000 and 10,500 radiocarbon years ago (or between about 24,000 and 12,000 calendar years ago) and have proven that the delta area was deeply depressed by the weight of the great ice sheets. John Clague, of Simon Fraser University, has found marine deposits near Vancouver that are now 200 metres above the present sea level. Given that sea levels during the height of the last glaciation were at least 100 metres lower than they are today, it seems clear that the isostatic depression in the lower mainland must have been at least 300 metres.

Once the great ice sheets were gone, the land rebounded quickly, much as a good mattress does when you get out of bed. Within 1,000 years, most of the lower mainland had returned to pre-glacial levels and the growing sediments in the delta were courtesy of glacial meltwater, and later, when the glaciers were gone from the province's interior, of the annual spring flood, the freshet. Together, these agents pre-served the remains of past cultures at dozens of sites.

Thanks to this ever-developing archive, and the geologists, paleontologists and archaeologists who have learned to decipher it, we now know that mammoths

Peter St. John

173

Whoever said of ocean-front property, "They're not making it any more", had clearly not been to Vancouver. All the land colored dark green has been added to the Lower Mainland in the past 10,000 years.

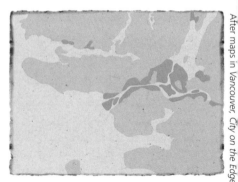

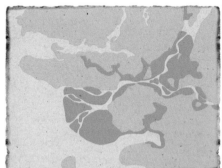

After maps in *Vancouver, City on the Edge*

inhabited the lower mainland as recently as 22,000 years ago, that people hunted elk and deer on the northwest slope of Surrey 8,000 years ago, that they gathered shellfish near Tsawwassen 4,500 years ago and that many of those who were buried at Crescent Beach on Boundary Bay suffered from painful arthritis. In short, thanks to the river delta, and the scientists who read its stories, we have an increasingly detailed picture of the long history of the great river and the rich cultures that thrived along its lower reaches.

We have also learned a good deal about the delta itself—not all of it reassuring—and we have at last begun to value this unique part of the world. Among the positives are these: the Fraser River estuary, which today drains more than 200,000 square kilometres of B.C., is among the richest in the world, the gateway to the largest salmon producing system on the Pacific Coast and a crucial international crossroad for migrating birds from twenty countries on three continents.

Upstream, where once the waters of the Strait of Georgia lapped at the shore of the Surrey uplands, is Burns Bog (see page 178), the largest raised peat bog on North America's west coast. Recently, just over half of this ecological jewel—only three per cent of the Earth's surface is covered with peatlands—has been set aside, saving it from development and further abuse and ensuring that its rare combination of plants and animals, and its crucial role as a natural air filter, will be preserved for the future.

On the less positive side is this; in a part of the world known for its tectonic complexity, the odds of a major quake in the next century are very high. And the Fraser delta—with its airport and sea port noted earlier—is highly susceptible to liquification. An earthquake registering a magnitude of 8 or more would almost certainly turn the delta's loose, sandy, water-saturated soil into something akin to quicksand, destroying bridges, highways, buildings and airport runways and cutting off many of the avenues needed to deliver emergency assistance.

To counter those threats, construction methods are improving, warning systems are ever more sophisticated and residents of the Lower Mainland are generally well informed about the potential for a major quake. Yet most seem unconcerned. Perhaps they believe that their daily exposure to the natural beauty and bounty of the Fraser delta is a fair trade for the risks it holds.

SFU Museum of Archaeology and Ethnology

Barbara Winter / SFU Museum

LOCATED ON Burnaby Mountain, across Burrard Inlet from Belcarra Regional Park (see page 196), the Museum of Archaeology and Ethnology at Simon Fraser University is less imposing, but more accessible, than its counterpart at UBC (see page 193).

Since the 1960s, when it plunged headlong into the province's distant past under the inspirational leadership of archaeology department founder and longtime chair Roy Carlson, the university has not only made huge inroads into the understanding of British Columbia's ancient past—but made that information widely available to an increasingly interested public. A great believer in accessible education, Carlson has been credited with inspiring the creation of the Archaeological Society of British Columbia and has mentored students of the discipline ranging from school groups and interested senior citizens to those pursuing doctorates in archaeology.

The museum's permanent exhibit features the excavation at Namu, on B.C.'s northwestern coast, one of the oldest archaeological sites thus far discovered in the province. But it also includes information and artifacts from sites from all over the province and indeed all over the globe. The university's focus on public education is also reflected in its on-line exhibits, which can be found at www.sfu.museum.

Among the thousands of artifacts in the university's collection are magnificently carved bowls, including the owl bowl, above.

Copper for the buttons in the centre of the page came all the way from the Copper River in Alaska.

175

The Glenrose Cannery Site

THOUGH LOCATED just a short distance from both the Delta Nature Reserve and its much larger extension, Burns Bog (see page 178), this is not a site that draws many visitors. Yet those interested in both Vancouver's geology and early human history will find the Glenrose site fascinating.

Walking the south shore of the Fraser River, across from Annacis Island, in the shadow of the Alex Fraser Bridge, a sharp-eyed beachcomber might find what archaeologists call a pebble chopper—a fist-sized stone with a sharp edge created by flaking shards from it with a hammerstone. And protruding from the beach are the stubs of waterlogged posts, which the dean of Vancouver archaeologists, Roy Carlson, believes are the remains of a fish weir. Radiocarbon dating has shown the posts to be 4,500 years old.

But much older remains have been found on the aptly named Panorama Ridge above the beach. Eight thousand years ago, people gathered every spring on the slope, just uphill from today's railway tracks and east of the cannery office. Then as now, Panorama Ridge was much higher than its surroundings. But people looking west would have seen not land, but water, as far as the eye could see. Panorama Ridge then was much like Point Grey is today, a peninsula extending into the Strait of Georgia, with the mouth of the Fraser River on one side, and a pebble beach below. As UBC archaeologist R.G. Matson has written, "The present Burns Bog area was probably a shallow bay, as the oldest dates for the post-marine deposits are slightly more than 5,000 years BP."

At the foot of the ridge, a stream ran into the water, attracting spawning fish in the spring. That, and the forested ridge above, made this an almost perfect site for some of British Columbia's early inhabitants. In fact, followng excavations in the 1970s, Matson believed that the site had been used by at least three cultures with very different lifestyles.

What has been termed the "Old Cordilleran" culture was the oldest and most diversified. In addition to fish from both the river and the sea, including salmon, flounder and eulachon, the people who lived by the river 8,000 years ago were accomplished hunters of both land and sea mammals. Excavating more than four metres below today's surface, archaeologists found four lovely leaf-shaped spear points, an unusual stemmed point, and knives, choppers, hammerstones and scrapers. Among them were the bones of the animals brought down with this array of weaponry—elk, deer, beavers and seals. Many were from young animals, adding to the evidence that this was a site used in the spring and early summer.

The artifacts found at Glenrose are similar to those found at several other sites on the B.C. coast, including one about the same age at Bear Cove on northern Vancouver Island.

Bear Cove was excavated in the late 1970s by archaeologist Catherine Carlson. The difference at Bear Cove was that nearly eighty per cent of the mammalian bones found belonged to sea mammals, mostly porpoises, but also including northern fur seals,

Peter St. John

Peter St. John

people were replaced, perhaps 4,200 years ago, by a new people almost wholly dependant on the sea. Carlson, however, believes that both the Glenrose and Bear Cove people were descendants of some of B.C.'s original inhabitants. Migrating down the coast at the end of the last glaciation, they had the technological ability to shift from hunting land and sea mammals to gathering clams and mussels as conditions changed.

Though their ancestry is still being debated, there is no question these "St. Mungo" people (the name comes from artifacts found at a nearby cannery site just to the west, which was destroyed when the Alex Fraser Bridge was built) had an elaborate tool kit. It included harpoon heads and fish-hook barbs made of antler or elk bone, along with finely wrought bone and antler pendants and even teeth drilled with holes, likely to be worn as ornaments. Wedges, crucial for people who worked extensively with cedar, were also found, along with scrapers, fleshers and other knives.

Above this layer of shells and artifacts was a third distinct cultural layer. For archaeologists, these remains were easily identifiable as belonging to sophisticated, elitist society known as the Marpole culture. Occupied between 2,500 and 1,500 years ago, Marpole sites are rife with tools of polished stone, ornaments of copper and carvings of bone and antler, a rich legacy of art, indicative, writes Roy Carlson, SFU professor emeritus, "of a well-developed ceremonial life".

The Alex Fraser Bridge soars over one of the oldest archaeological sites on the Lower Mainland, creating an incongruous mix of present and past.

Stellar sea lions, sea otters and harbor seals. Clearly, the Bear Cove people were very much at home on the water and must have had, writes Catherine Carlson, both well-made watercraft and efficient harpoons.

Carlson believes that Glenrose and Bear Cove are alike in another way; they were both home to accomplished hunters and gatherers who were accustomed to a seasonal round that included hunting, fishing and gathering shellfish at appropriate times of the year.

At Glenrose, the upper levels of the excavation contained dense shell middens. Matson believes this indicates that the "Old Cordilleran"

One of the largest such sites once lay across the river and downstream from the Glenrose Cannery. Located on what is now Marine Drive across from the east end of Sea Island, it covered nearly two hectares and was at least five metres deep; the Fraser Arms Hotel sits at what was once the centre of the site. Not surprisingly, it was known as the Great Fraser Midden (see page 186).

Occupied for at least a thousand years before the Fraser Delta reached its present size, it marked an apex of the Marpole culture and established the reputation of at least one early archaeologist—American Harlan I. Smith.

Getting There: The Glenrose Cannery Site is located north of River Road and east of the Alex Fraser Bridge just above the railway tracks.

Peter St. John

Burns Bog By Philip Torrens

TRAVELLING BACK IN TIME to see the world as it once was is a common daydream for children and adults alike. Burns Bog offers perhaps the closest thing to that fantasy that's available in Vancouver today—the opportunity to see an environment largely unchanged for 3,000 years. And you don't have to travel to the ends of the Earth to experience it—this domed bog, the largest on North America's west coast, sits just twenty kilometres from the city centre, and is remarkably accessible.

For many, the words swamp, marsh and bog are interchangeable. But to specialists, they denote distinct environments. Swamps are wetlands that may be treed or tree-free; marshes are reedy wetlands that are unforested; a bog is a wetland that may or may not have trees, but it must have one thing the other two wetlands lack: moss—and lots of it.

It's moss that makes a domed bog a self-assembling, self-perpetuating and self-propagating organic island amid the surrounding ecosystems. Many islands emerge as the result of falling sea levels, rising land, deposited silts, or volcanic eruption, and are colonized by living things only after they've formed. But domed bogs, like coral islands, are built by living things atop the remains of other things.

Fifteen thousand years ago, glaciers covered much of B.C. to a depth of thousands of feet. When the ice retreated, it left behind many deep depressions gouged into the land, including a trench that became the incubator for Burns Bog. This particular hollow, near what was then the mouth of the Fraser River, filled with a blend of fresh water from the river and brine from the Strait of Georgia to become a wetland. As the river grew, it formed a delta that eventually dammed off the area. In the isolated and now stagnant waters, logs that had not quite been swept to the

sea intermingled with cattails, rushes, grasses and sedges. All these plaited into a thick layer of rotting plant matter as they died and fell. With little water drainage to flush away the acids generated by this decay, the environment became hostile to many plants, but ideal for sphagnum moss, which soon took hold in a thick carpet.

Able to retain dozens of times its own weight in water, the moss grew upward, layer upon layer, at perhaps a millimetre a year, like a giant, slow-growing sponge. (Sphagnum grows so slowly that its roots are actually rotting while the tips are thriving.) And it grows still; currently, the center of the bog's dome, and its water level, rise a full six metres above the nearby Fraser River.

Beneath the layer of living moss, the dead plant matter is trapped in a low-oxygen, acid environment that inhibits complete decay. Instead, the vegetation becomes semi-carbonized, forming peat (sphagnum is also known as peat moss). Peat will eventually become coal, given the proper temperatures and pressures and several hundred million years. Perhaps lacking that kind of patience, many cultures around the world have used peat for fuel; it burns well, and unlike coal, if carefully harvested in narrow swathes, will renew itself.

Peat burns so readily that fires

© David Blevins (top & bottom)

have spread through Burns Bog on several occasions, most recently in the summer of 2005. This fire generated a plume of smoke that could be seen for a great distance. Even downtown Vancouver was under a peaty pawl for several days, reminiscent of Scottish hearths perhaps, but not pleasant for those with respiratory problems.

The fire smouldered for many weeks before it could be extinguished, presenting novel challenges. Simply finding the fire's boundaries can be tough; one solution is to employ infrared imagers, which can detect the heat of a blaze. Firefighters on the ground must be particularly careful, for pits of fire can go undetected and a tunnelling blaze can entrap them.

This was neither the first nor the longest -burning fire in living memory: major fires occurred in 1977, 1990, 1994 and 1996. During World War II, when peat was harvested for magnesium for use in manufacturing munitions, a fire burned for four years.

Far from being invariably fatal for bogs, fires can actually help preserve them. The pine trees infiltrating along the bog's borders "exhale" moisture into the air much faster than peat does. This lowers the water table. If not halted, the forest will eventually drink the water level down below the

© David Blevins

point of no return for the peat and permanently displace the bog. By periodically beating back this encroachment of trees, fire serves as the bog's natural defender. Provided the water table is preserved, the moss will regenerate in the wake of a fire.

In addition to its flammability, sphagnum moss has another interesting property: it contains a natural antibiotic called tropolene. Native peoples used dried peat moss for bedding, diapers and feminine hygiene pads. Europeans used it to dress wounds during World War I.

Because they make such effective medicines, we tend to think of antibiotics as invariably good things, but the word antibiotic actually means "anti-life." And in fact, the presence of sphagnum moss and its accompanying tropolene creates a very challenging environment for other life forms; trees decades old may be only a few feet high. But those life forms that can rise to the challenge of sharing space with sphagnum moss enjoy the opportunity to ply their trades amid much reduced competition.

Though it has an excellent pair of natural defences in the form of fire and tropolene, there is one threat the bog cannot protect itself against: human activity. Only people can protect the bog against people. Until 2004, of the bog's approximately 3200 hectares, just twenty-four were protected as the Delta Nature Reserve. That year, the three levels of government acted in concert to purchase a further 2023 hectares. Protected as an Ecological Conservancy Area, the intent is to emphasize preservation over human access. The Burns Bog

SOUTHERN COMFORT

Besides being an oasis in time, Burns Bog is an oasis in space: many of the life-forms found here are usually seen only in sub-arctic or alpine environments, instead of the temperate lowlands of the Fraser Delta. Here are just two of its exotic inhabitants: the spicy-smelling white flowers of Labrador tea (shown here) invite one to brew it up. Tea can be made from either fresh or dried leaves. Many cultures in B.C. used Labrador tea as a treatment for sore throats and colds. Early European settlers brewed it as an everyday beverage, apparently because it was relaxing. That relaxation may come from small doses of the poisonous alkaloids in the leaves. The tea must be drunk carefully and in moderation; excessive consumption can cause severe intestinal upset and dehydration, and some people are more sensitive than others to alkaloids.

The round-leaved sundew is a relative of the Venus flytrap. Sundew leaves bristle with spiky red tentacles, giving the plant the appearance of a miniature sci-fi movie monster. And indeed, the sundew is highly lethal to any mosquito, midge or gnat unfortunate enough to be trapped in the sticky "dew" that coats its leaves. These insects will eventually be dissolved to stoke the sundew with nitrogen, an essential nutrient for plants and one in short supply in the bog. Coastal peoples used sundew leaves to remove warts, bunions and corns.

© David Blevi

Conservation Society, a 5,000-member grass roots group, also hopes to purchase additional land from a private owner. The more land protected, the better the chance of preserving a critical mass that will allow the bog to sustain itself.

Nor is this dependency for health a one-way street. The iron-rich runoff from the bog is vital to maintaining fish populations. Even more significantly, bogs store up to three times more carbon dioxide than rain forests of the same area. The destruction of bogs releases vast amounts of carbon dioxide into the atmosphere. This transparent gas acts like the glass of a greenhouse, letting the sun's rays pass easily through, but resisting the escape of the heat from Earth's sun-warmed surfaces. This greenhouse effect contributes to global warming. We often wonder what we should do about global warming. One important answer is, do nothing—nothing to destroy the carbon dioxide trapped in bogs and forests.

Getting There:

By car Take the River Road/Nordel Way exit from Hwy 91. Follow the signs for River Road, onto Nordel Way, heading north. Turn right at the second traffic lights onto Nordel Court. Continue to the Great Pacific Forum (10388 Nordel Court). Park in the southeast corner of the parking lot. From here, a red brick path leads to the left and under the Nordel Way overpass. Follow the gravel GVRD access road running south into the Delta Nature Reserve.

By public transit Take the Skytrain to the Scott Road Station. From there, ride the 640 Tsawwassen bus to the Sidetrack Pub on River Road. Follow the GVRD access road south.

Crescent Beach

CHILDREN PLAY IN THE SAND and families walk the tidal shallows of Crescent Beach, which stretches south from Mud Bay along the northeast corner of Boundary Bay. Out on the water, small boats ride the incoming tide. It's easy to imagine a similar scene almost any spring day during the past 5,000 years, for Crescent Beach has a long and fruitful history. Here, in large open clay ovens measuring a half-metre or more across, clams and mussels were steamed and shelled, while ever-abundant herring were roasted and dried over nearby fires and herring spawn was dried on high racks, well out of the reach of animals.

The shellfish ovens, constructed of thick clay directly on the beach sands, were used year after year. So durable were they that several were still intact when, more than 4,000 years after their creation, archaeologists excavated one of the deep shellfish mounds in the early 1970s.

Their reports, along with artist Gordon Miller's lovely painting, below, make it easy to imagine this scene say, 2,000 years ago:

A blazing fire on the beach heats rocks. Pausing as they shuck steamed mussels and clams, the women carefully place red hot rocks in a pit to steam another batch of shellfish. Layering the clams and mussels

© Gordon Miller

with seaweed to provide the moisture needed to make steam, they cover the pit to keep the heat in.

Children romp in the shallows and just offshore, teams of two—many of them husband and wife—are harvesting herring spawn.

The waters of the Strait of Georgia have been all but empty of herring for more than a half-century, but for millennia, these tiny fish were so abundant in early spring that they turned the dark waters of the strait to silver as they streamed toward the shallows to spawn. There, on waving stands of kelp—or on weighted fir, cedar or hemlock boughs that many coastal cultures hung on frames to catch the herring spawn—the females laid their eggs in the millions. Gathered at low tide after the runs were finished, the egg-covered branches were collected and hung to dry. Many communities around the Strait of Georgia collected and dried herring spawn, often trading this delicacy inland. Even today, some native British Columbians collect herring spawn.

But it may not have been just shellfish and herring that were processed at Crescent Beach, according to a report by Richard Percy, formerly curator of the Museum of Archaeology and Ethnology at Simon Fraser University. As early as 4,000 BP, the site may also have been used for manufacturing chewing tobacco. This was created by combining native tobacco, which once grew north of Crescent Beach, with finely powdered lime from the shells that were heaped in piles following each year's processing.

Not surprisingly, given the long occupation of the site, an enormous number of artifacts were found when this section of Crescent Beach was excavated prior to the installation of sewer lines in 1972. Most of the artifacts, about 1,500 knives, scrapers, sinkers, bowls, mauls, chisels, awls, harpoons, beads and many other objects were made of stone. But some were of bone, antler and teeth and a few were of shell, including beautifully crafted bowls, beads, pendants and labrets.

But perhaps the most magnificent find was not uncovered until three years later, when Percy discovered that the soil excavated from the site during construction had been dumped in the parking lot of a nearby

marina. With members of the British Columbia Archaeological Society, he decided to see whether there was anything left to recover. Working weekends, the team found about 250 additional artifacts; the prize was a huge, superbly worked biface or spear point that measured more than a third of a metre in length. The magnificent point, as well as a labret, were found by the same veteran volunteer—"Lucky" Jim Garrison—within ten minutes of one another. Both are now in the collection at SFU.

The artifacts and the excavation told the same story; Crescent Beach was used seasonally, almost continuously, for nearly 5,000 years, very likely

THE SEMIAHMOO TRAIL
This trail was an ancient Indian travel-way linking tribal villages in the south to salmon grounds of the Fraser River. The first white explorers, led

by the Halq'emeylem-speaking people who appear to have occupied the region almost until historic times. After they were devastated by smallpox, which often preceded the actual arrival of Europeans, the area was annexed by the Semiahmoo people, from just south of what is now the U.S. border, and closely related to the Lekwungen and T'Sou-ke across the Strait of Georgia.

Getting There: Take Hwy 99 south from Vancouver, exit just before Hwy 99 intersects with Hwy 99a. Follow the signs to Crescent Road and follow Crescent Road to its end.

The Beach Grove Site

SURPRISINGLY PERHAPS, the Beach Grove Site is not on the beach. Located both north and south of the junction of 16th Avenue and 56th Street in Tsawwassen, it lies well inland from the shoreline of today's Boundary Bay. But the name of the site in both English and Halq'emeylem (or Tsawwassen), "looking toward the sea", indicates it was not always thus. No fewer than ten excavations over the past half-century have allowed scientists to develop a timeline for environmental change in the region, and have demonstrated how the people who once lived here adapted over a period of almost four millennia, between 4,500 and 600 years ago.

Tsawwassen is the Canadian portion of a peninsula that lies across the 49th parallel; its southern extremity, an isolated chunk of American territory, is called Point Roberts. Today, Tsawwassen is part of Greater Vancouver, linked to the city by the lowlands of Ladner and Richmond and is probably best known for its ferry terminal. But Tsawwassen was not always part of the mainland. In fact, for most of the past 12,000 years, it was one of the Gulf Islands; geologists still call it Roberts Island.

It emerged from the waves of the Strait of Georgia when the post-glacial sea level, which had sharply risen as a result of the enormous volumes of meltwater, receded to about seventy-five metres higher than it is today. Over the next 3,000 years, as the sea level dropped to about twelve metres lower than modern levels, Roberts Island grew ever larger. In places, such as the southeastern tip, it might have extended into the strait as much as a kilometre farther than it does today.

After stabilizing for about 2,000 years, the sea began to rise again, slowly encroaching on the island. The southeastern tip slipped under the ocean, and became the region's most bountiful reef netting site (see page 238). Meanwhile, the water and wind began the lengthy process of moving glacial silt and sand from the ancient shorelines up the western and eastern shores. By about 6,000 BP, a sandy beach ridge— dubbed Pillars Inn Ridge for a motor inn that would be located on the site thousands of years later—had begun to build into the strait at the island's northeast tip. Some time later, a second ridge developed along the north shore slightly to the west. When the two ridges joined (see map), perhaps 3,500 years ago, they created a triangular lagoon between them. This lagoon slowly filled with vegetation over the next thousand years until, about 2,300 years ago, it had become a peat bog. To the east, the Fraser River delta was building the undersea bridges that would one day support the natural link to the mainland.

It's possible—even likely—that there were people living on Roberts Island 6,000 or 8,000 years ago. But

Peter St. John

Today, the beach is several blocks from the Beach Grove Site, demonstrating the power of wind, water … and time.

Peter St. John

Part of the original site lies under the Tsawwassen's Beach Grove Golf Course, providing members with more than golf scores to discuss at the 19th hole.

it wasn't until about 4,500 BP that a shell midden—a refuse area that archaeologists use as a clear indication of regular or continuous settlement—began to develop on Pillar's Inn Ridge. (The second ridge, dubbed McDonald's Ridge, was never a favored campsite, probably because it was exposed to the full force of the prevailing northwesterly winds.)

Pillar's Inn Ridge, on the other hand, was apparently a perfect place to live for millennia. Though archaeological evidence shows that, over time, its inhabitants had to make significant changes to accommodate the evolving environmental conditions, the midden continued to grow and spread for nearly 4,000 years, ultimately covering an area of nearly 200,000 square metres, including much of what is now the Beach Grove Golf Course.

Shells recovered from the lowest levels of the midden show that 4,500 years ago, the settlement lay on the western side of Pillar's Inn Ridge, which was then directly on the sea, but protected perhaps by the emerging McDonald Ridge to the northwest. When that ridge eventually cut off the community's sea access, the people did the obvious: they moved across the ridge to the eastern shore, facing what is today

Boundary Bay. At the time, however, this was still an island, separated from the encroaching mainland by a tidal channel. Archaeologists have found beach ridges on top of the delta deposits, and have dated the midden deposits from the eastern side of the beach ridge to about 3,470 BP. It seems that the delta finally reached the island, creating Boundary Bay, about 3,000 years ago.

The land bridge was not all good news; while it gave access to the mainland, it also disturbed the intertidal flats that were once found all along the northeast shore of the island. Sand and silt began to build up, creating an area known today as Beach Grove. As the centuries passed, the people had to travel ever farther east to reach the sea, with its abundant shellfish, fish and sea mammals. Nevertheless, it's clear that the community thrived. And artifacts recovered over the past twenty years show that it was not only in touch with artistic and cultural developments on the mainland, but had developed its own intricate and decorative techniques for such things as cordage and basketry. In 1988, when one of the golf course's water links was enlarged, construction crews dredged up mud containing a wealth of perishable artifacts. For the next month, archaeologists, members of the Tsawwassen and Musqueam communities and local volunteers dug through the mud, which was kept wet with sprinklers. In the end, they recovered more than 300 artifacts—segments of intricately twined cedar baskets, handles and cords, as well as fish hooks of Douglas-fir and wedges made of yew. All dated to between about 2,000 and 1,550 years ago, a sophisticated cultural era that archaeologists call the Marpole period (see page 186).

About 1,000 years ago, the village was once again moved east, to the new shore of Boundary Bay. For a time, parts of the old Beach Grove Site continued to be used as a burial ground, but about 600 years ago, Pillar's Inn Ridge was finally abandoned.

The last in the long series of excavations, which was undertaken in early 1995 with the assistance of the Tsawwassen and Semiahmoo First Nations, unveiled this remarkably adaptable community. As archaeologist Richard Brolly has written: "As the original marine nearshore environment of Roberts Island evolved first into an estuarine setting, and later still into a landlocked delta lowland, the … occupants of the site adapted to these changes."

It should therefore come as no surprise that the descendants of the Beach Grove Site have also proven to be remarkably adaptable. Today, land owned by the Tsawwassen people is leased for private homes, a waterpark, ferry parking and a condominium development. And artifacts found during the excavations are on display at the Delta Museum.

Getting There: The original Pillar's Inn Ridge site has been largely developed, but part of the main Beach Grove site lies under the golf course of the same name just south of 16th Avenue. Beach Grove, where people now walk their dogs, lies just to the east at the end of 16th, while Boundary Bay Regional Park encloses the area between the original eastern shoreline of Roberts Island (on the west side of the park, along Boundary Bay Road) and the Boundary Bay and Beach Grove Spits.

The Dyke Trail, which winds along the eastern shore of the park and joins the Boundary Bay Trail along the northern shore of the bay, and the Raptor Trail, which winds through the middle of the Wildlife Reserve, offer a glimpse of the past. Parking can be found off Boundary Bay Road, which is accessed off 56th Street, south of Hwy 17 (the main road to the ferry).

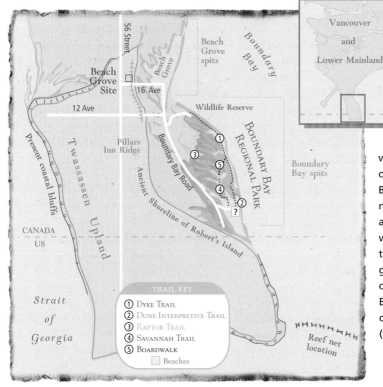

TRAIL KEY
① Dyke Trail
② Dune Interpretive Trail
③ Raptor Trail
④ Savannah Trail
⑤ Boardwalk
☐ Beaches

The Great Fraser Midden

TODAY, THE LAND at the foot of Granville Street seems rather ordinary, with its city streets and sidewalks and a hotel parking lot. Freeway traffic hums nearby. The Oak Street Bridge looms above and the view to the west, over the North Arm of the Fraser River, is of Sea Island with its planes taking off and landing at Vancouver International Airport.

But this site is anything but ordinary. It might, in fact, be called Vancouver's buried treasure, for on the slope above the river, beneath the asphalt and the parked cars, are the remains of a people whose art and culture inspired communities throughout the Gulf of Georgia for more than a thousand years. This is the Great Fraser Midden or, as archaeologists often call it, the Marpole Site. Stretching over two hectares, piled to a depth of five metres, the midden goes back at least 3,000 years, to a time when Ramses III reigned over Egypt, and Troy was being sacked.

The ancestors of the Coast Salish people who occupied the Marpole site had lived along the shores of the Fraser River since the glaciers retreated. For them, this was Sqla-lot-sis, "the homeland". And it's almost certain that people had fished and hunted right here, where the west-trending shore of the delta angles gently northwest into the Strait of Georgia, long before it became a

A particular type of unilaterally barbed harpoon points (top and left) were unique to Marpole culture, allowing it to be tracked from the San Juan and Gulf Islands to the mouth of the Fraser (opposite, top). Leisters, three-pronged harpoons (opposite), were also widely used.

Images this page courtesy of Roy Carlson

centre of arts and culture, for this site—and the river itself—were very different 5,000 years ago than they are today. Then, looking south from the shore, instead of low-lying islands, one would have seen the Gulf of Georgia stretching almost as far as the eye could see. Perhaps, in the far distance across the water, the forests of Roberts Island—as geologists call it, or Tsawwassen, as we know it today—could be seen, hovering on the horizon. To the east, what would one day be Lulu Island—the site of today's Richmond—was growing in the river delta and the swampy lowlands of Delta and Ladner were stretching, like an out-stretched hand, from the heights of Surrey, to bridge the channel that separated them from Roberts Island.

The Marpole site was therefore perfectly positioned to harvest the growing numbers of salmon that swarmed east each year into the river's mouth, heading for tributaries far upstream where they would spawn—or die trying. Salmon were key to the creation of the sophisticated community that would rise here, and to the outpouring of art and culture it nourished. Six salmon species—sockeye, pink, chum, chinook, coho and steelhead—passed through the estuary. Together, they provided a dependable, readily obtained, easily preserved source of food that largely freed the people from the daily necessities of hunting and gathering. By the time the village at the Marpole site was established, tens

Sir John Douglas Sutherland Campbell, National Archives of Canada, C–111834

of millions of salmon returned annually to spawn, making the river the world's largest salmon producer. This bounty not only sustained the villagers, but dried or smoked, could be traded as well.

The Fraser harbored more than just salmon, of course. Upwards of eighty species of fish and shellfish made their homes in the estuary's many habitats—from its bountiful kelp and eelgrass beds to its freshwater marshes and sand and mud flats. Among them were huge sturgeon, which returned to the river each spring to spawn, and herring and eulachon that filled the bays of the Straits of Georgia in March and April, pursued by voracious congregations of whales, seals and sea lions. On the mud and sand bars, shellfish proliferated in such numbers that the mound of discarded shells ultimately grew deeper than the roofs of the people's magnificent cedar homes were high. And this aquatic plenitude drew millions of migrating shorebirds, waterfowl and gulls.

Behind the village, the climax forest was filled with towering Sitka spruce, Douglas-fir, western redcedar and hemlock which, over time, encouraged a woodworking industry that would one day be synonymous with British Columbia itself. Little wonder that each summer the Fraser delta drew visitors from far and wide, to fish, trade, exchange ideas and, undoubtedly, find husbands and wives.

This enormous bounty not only meant considerable wealth but, about 2,400 years ago, the flowering of a complex culture with the leisure to indulge in personal adornment and sophisticated art and woodworking.

Both the size of its midden and the art and tools it contained set the Marpole site apart almost as soon as lumberjacks and construction workers stumbled across it as they began work on an extension of Granville Street in 1889. In the first decade, archaeologists discovered exquisite stone bowls in the shape of seated human figures; fish, seal and turtle effigies; pendants of antler; bone and antler needles and awls, and ornaments and beads of shell, shale and even copper —though the closest known source is the Copper River in Alaska. They also discovered more utilitarian objects, including bone needles and awls, chipped stone projectile points and wedges, mauls and celts that were clearly used for large scale woodworking. But unique to Marpole society were unilaterally barbed harpoon points made of antler. It was these that would allow the spread of Marpole culture to be tracked throughout the region.

Archaeology was in its infancy when the site was uncovered, but soon after Franz Boas of the American Museum of Natural History arrived in 1897 for the first of several years of excavating, two things became clear. The depth of the midden suggested remarkable cultural longevity and the magnificent artifacts made it clear that this was a society of wealth and power. As well, some of the disinterred skeletons were found to have deformed skulls. Boaz believed the skulls meant the presence of a new people on the Northwest Coast, but he also reported that the Marpole site clearly held the roots of the rich Coast Salish culture that greeted the first Europeans to the lower Fraser.

In fact, anthropologists are now quite certain the cranial deformities found at the Marpole site likely had more to do with ideas of style and beauty than with the arrival of a new people. It's likely that the heads of infants were sometimes bound to create

187

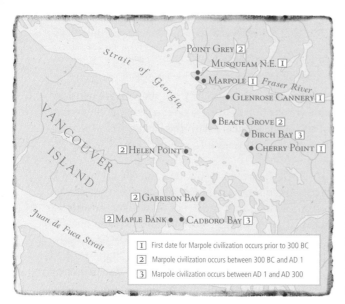

Map legend:
1   First date for Marpole civilization occurs prior to 300 BC
2   Marpole civilization occurs between 300 BC and AD 1
3   Marpole civilization occurs between AD 1 and AD 300

a wide, flat forehead that was not only considered beautiful, but may have served to set certain individuals or families apart. There were many other indications that fashion was important. Shell and stone labrets and earspools, rather like today's lip and ear rings, were sometimes worn, and pierced ears were common.

In the past fifty years, work by Carl Borden, Roy Carlson, David Burley and others has shown that Marpole was more than simply one remarkable site; it has become what experts call a "type site", the place of origin of a culture that is widely mimicked. Excavations elsewhere in the Vancouver area, the Gulf Islands and Vancouver Island have shown that Marpole culture was widely adopted over the 600 years between 2,300 and 1,700 BP. From Millard Creek near Courtney on Vancouver Island's east coast, southwest to what are now Victoria's western suburbs and southeast to several sites in today's Washington State, community after community echoed the art, technology, cooperative style and organization of the people of the Marpole site. The discovery of large, square post moulds and the outlines of houses at a number of sites, including the Beach Grove and False Narrows, make it likely that the kind of large, often elaborately decorated, multi-family cedar plank homes that greeted the arrival of Europeans had their origins at Marpole.

All these things suggest that Marpole culture was complex and well organized and scientists from several disciplines have developed theories about how and why this remarkably successful society evolved. First, though it does not seem to have been a highly stratified society, as was Keatley Creek (see page 143), for example, archaeologists believe that life at Marpole would have required the leadership of citizens or families, as well as specialists in many areas—artists, weavers, woodworkers and basket and tool makers, in addition to the fishers, hunters and gatherers who sustained the growing community. Maintaining their place in the centre of this bountiful region would have required the people of Marpole to become diplomats, to train citizens who were widely respected for their wisdom, judgement and, perhaps, generosity. The last of these may have led to the development of potlatches, the large-scale ceremonies that came to symbolize many cultures up and down the Northwest Coast. These gatherings not only celebrated the erection of new houses (marvels of intense labor and cooperation), marriages, funerals and the transfer of hereditary rights and titles, but also served as a means for distributing wealth, assisting those in need while reaffirming the donor's social status.

Though it's more than a century since the Marpole site was first excavated, the interest in its people and its culture continues, changing as our knowledge in many areas expands. In the April 2005 issue of *American Antiquity*, an article by a quartet of climate specialists proposed that the social and economic networks—and perhaps by extension the dominant nature of Marpole culture—may have been sparked by climate change. Though it was long thought that the period between 2,400 and 1,200 BP was one of cooler temperatures, new information seems to indicate that the opposite was true. The authors contend that this was, in fact, a period of global warming that they call the "Fraser Valley Fire Period". Characterized by persistent summer drought and a significant increase in the number of forest fires, they believe this longterm climate change would have led to a regional decline in salmon—accompanied by a greater abundance of both berries and wildlife. Over time, summer drought conditions would also cause the thick undergrowth of the rain forest to thin, allowing greater overland access and perhaps encouraging the creation and increased use of overland trails.

The authors hypothesize that all these factors strengthened the social and economic networks throughout the Gulf of Georgia, as people inland and in drought-stricken areas sought to ensure access to the resources of the Fraser Valley. And this, in turn, increased the stature and prestige of those living along the shores of the lower Fraser River. Certainly, decorative and ceremonial items—dentalia shells, mother-of-pearl, copper and jade—were all traded widely during this period, as were beautifully woven baskets and magnificent blankets.

Interestingly however, what the authors indicate as the end of this period—about 1,200 BP or 800 AD—marks the beginning of both a period of global warming that European historians call the Medieval Warm Period (known to have greatly affected the North America's continental interior) and centuries of what appear to be increased hostilities in the Gulf of Georgia (see Finlayson Point on page 236).

A thousand years ago, other things were also changing at the Marpole site. Extended year by year by the river's load of sand and silt, Lulu Island was growing and Sea Island was forming in the wide expanse of water in front of the ancient community. Over time, this changed the spawning patterns of the salmon and turned the salt water flats to freshwater beaches.

Finally, the people abandoned their cherished home and moved downstream to the neighboring Musqueam villages, whose linked cedar houses, facing the water, were undoubtedly the huge "fort" that Simon Fraser saw on his arrival at the mouth of the river in 1808. But by that time, greatly diminished by disease and out of patience with European intruders, the Musqueam turned on Fraser's small crew, likely feeling that the best defence was a good offence.

Getting There: The Great Fraser Midden is buried beneath the asphalt and concrete of the Fraser Arms Hotel at 1450 SW Marine Drive, though that may change. Recently, Parks Canada has put the Marpole Midden on its list of National Historic Sites, and the Musqueam people, who own a significant tract of land along the river west of the site, have purchased the hotel and parking lot, with the intention of creating an interpretive centre. For the time being, a stone cairn and plaque in Marpole Park on recalls one of North America's most significant cultures. The Museum of Anthropology at the University of British Columbia and the Museum of Archaeology and Ethnology at Simon Fraser University have artifacts from the magnificent Marpole culture on display, while the Vancouver Museum has artifacts from the Great Fraser Midden in its collection, though not on display.

Peter St. John

189

VanDusen Botanical Garden

FROM VANCOUVER ISLAND'S endangered Garry oak woodlands (see page 240) and the ponderosa pines of the equally threatened shrub-steppe ecosystem of the southernmost Okanagan (see page 126) to the towering Douglas-firs of the Gulf Coast, British Columbia is Canada's most biologically diverse province. Yet many of its species are under threat from land development and exploitation of resources, from the invasion of exotic species and diseases and from global climate change.

As part of a worldwide effort to encourage plant diversity and create live gene banks, Vancouver's beautiful VanDusen Botanical Garden is increasingly focusing its efforts on preserving and showcasing British Columbia's native plants and ecosystems. It is, in short, becoming a living museum.

Not surprisingly, perhaps, some of species and ecosystems most endangered are those that are native to the province's most heavily populated areas. Garry oaks, for example, have quite literally survived mainly because they are grown in urban backyards and city parks in Victoria and Vancouver, while the rare plants of the antelope-brush desert continue to exist in large part because of the stewardship of both the Okanagan First Nations and the concerned citizens of Osoyoos of the southern Okanagan.

Examples of these plants or ecosystems can all be found at VanDusen (along with magnificent plant collections from five other continents), but new efforts are being made to expand, preserve, protect and promote the garden's native plants. The garden's arbutus trees are a good example. With their twisting, peeling, cinnamon-barked trunks and large evergreen leaves, arbutus—Canada's only native broad-leafed evergreen that is also known as the Pacific madrone—is prone to at least nineteen different fungi that cause leaf spots. Most of these develop just before the leaves are shed, but gardener Egan Davis, who works on VanDusen's North American garden, says more worrying is the recent widespread discoloration and premature dropping of leaves that has been noticed throughout the Lower Mainland, the Gulf Islands and southern Vancouver Island. Davis and others believe that stresses, including drought and air pollution, may be the cause.

While examples of native plants can be found in a number of places throughout the 22.4-hectare botanical garden, many are concentrated along the stream northeast of the garden's excellent Shaughnessy Restaurant. Plants are often identified with markers, the gardeners are always happy to assist and there is a reference library on site, as well as a gift shop that carries many excellent reference books on B.C. plants.

Getting There: VanDusen Botanical Garden is located at 5251 Oak Street (near the corner of Oak and 37ᵗʰ) in Vancouver. Open year-round, it is beautiful in any month of the year.

Allan Taylor

Arbutus bark and leaves were brewed to treat colds and other illnesses.

Point Grey

THOUGH THE HERRING are all but gone and the surrounding forest lacks its primeval majesty, it's still possible for visitors to Point Grey to capture the essence of the place as it was 2,000 years ago.

On a raised bluff on the northwest tip of the point, twenty metres above the beaches of Spanish Banks and just west of Spanish Banks Creek, is a Marpole spring campsite that was occupied repeatedly over a period of at least 600 years. Beginning about 2,200 years ago, the ancestors of the Coast Salish people arrived here every year in late February or early March, just as the Pacific herring began to spawn.

Its almost unimaginable now, looking out over the quiet water, but for many millennia the Strait of Georgia and Burrard Inlet seethed with life. The spawning of the herring turned the surface of the water a silvery black and filled the air with a dull roar as millions of fish flooded the inlet. Balls of herring bulged from the water as pods of minke whales pursued them and flocks of shearwaters dived among them. Coming here meant a raucous end to winter for people who undoubtedly were ready for something other than their by-now monotonous diet of dried salmon.

Marpole culture was a complex, stratified, "ranked" society, marked by differences in social status, sophisticated art and craftsmanship and a tendency to live in large, permanent or semi-permanent villages. The Point Grey site was not, according to the excavations done in the past twenty years, one of those villages. Rather, it appears to have been a spring campsite, a place where large groups came on a regular basis to fish and gather shellfish, including littleneck and butter clams, as well as mussels, oysters, cockles and whelks.

Though the shellfish were plentiful on the beaches and rocky shores, and gathered in such enormous numbers that the shell debris eventually covered an area eighty metres wide and nearly 300 metres long, it appears that fishing was the primary spring activity and herring by far the species of choice. These footlong fish made up more than ninety-five per cent of the catch; other fish taken, like salmon, were probably incidental quarry, likely following the same schools of herring the people were.

Schooling in numbers that are almost unimaginable today and intent on spawning, herring were easily caught with dip nets or herring rakes, long poles with hardwood or bone teeth set—comblike—into holes at one end. With one person paddling, another swept the rake against the current, just under the surface of the water. Loaded with wriggling fish, it was lifted and swung toward the canoe. A sharp rap on the gunnel and the catch was added to a growing pile in the belly of the boat. Later, threaded onto a pole and baked over a fire, these small fish were the taste of spring.

Nor were the herring the only catch; the fishers were also after their eggs, or spawn, considered a delicacy all up and down the West Coast. Beginning in early March, the fish began to congregate along the coast in quiet bays and coves, where they deposited their eggs on what *Georgia Straight* writer Nancy Baron has lyrically termed "forests of kelp and plains of seagrasses."

Offering an alternative to these, the Marpole people built rectangular log frames, and anchored them in the bays preferred by the fish. Spaced so that canoes could be paddled between them, the frames

Herring not only drew people to Point Grey every spring, but also hungry sea mammals and flocks of birds.

Dawn Huck

191

A log herring frame, grounded by a stone anchor, awaits the spawning fish.

Dawn Huck, after drawing in *Indian Fishing*, by Hilary Stewart

Getting There: Both 41st and 16th Avenues lead into Pacific Spirit Park and both are bus routes. The Park Centre is located on 16th, just south of the Blanca Bus Loop. The park is crisscrossed by trails and paths, some (including the trail around the perimeter of the point) are for pedestrians only, others are multi-use trails for bikes and horses as well as walkers. Organized nature walks take place year-round.

held fir boughs, tied to the logs and weighted with stones to hold them underwater. Moving gently like seaweed, the fir boughs attracted the spawning herring, which covered the branches with eggs. The resulting cornucopia was eaten fresh or dried.

Though archaeologists excavating the Point Grey spring village found few of them—likely because this was a seasonal working camp—the Marpole culture is widely known for its exceptional ornamentation; even utilitarian objects, such as harpoons and floats were often wonderfully carved.

Today, the Point Grey site lies at the north edge of part of the 763-hectare Pacific Spirit Regional Park, which was established in 1989 on the forest endowment lands that lie between Vancouver and the University of British Columbia. Stretching across

Point Grey from Burrard Inlet to the Fraser River, and encompassing a narrow shoreline margin around the edge of the point, the park allows visitors a glimpse of yesterday (see map on adjacent page). Estuary marshes, wooded ravines, huge cedars and Douglas-firs are all reminders of what Point Grey was like during the Marpole era. We can only hope there is still time for herring moratoria, a new interest in the wellbeing of the great whales and quotas on salmon and other species to reverse the dreadful decline in British Columbia's oceans, and give us a glimpse of the sea that the Marpole people knew so well.

Peter St. John

Instead of canoes and herring fishers, today Point Grey looks out on tankers and ships on the waters of Burrard Inlet.

Vancouver Museum of Anthropology

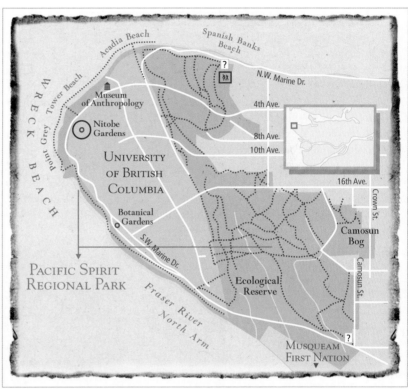

WEST OF THE POINT GREY SITE, on the northwest tip of the peninsula, on a site the X'muzk'i'um or Musqueam people sensibly called *ʔelqsen* or "point", is Vancouver's striking Museum of Anthropology.

Founded more than a half-century ago in the basement of the University of British Columbia's main library, the museum has developed into B.C.'s single most comprehensive collection of Northwest Coast art, artifacts and archaeology. Among the 535,000 objects are monumental totem poles and beautiful bentwood boxes, as well as masks, carvings and blankets.

Designed by celebrated Canadian architect Arthur Ericson in 1976, the museum building, and its grounds, are as worthy of a visit as the rotating exhibits, research collections—including many in "visible storage"—library and gift shop. Inside and out, Erickson's design evokes the massive post and beam structures of the Northwest Coast, and the focus of its permanent exhibits is very much on the peoples of the region—the Haida, Gitxsan, Kwakwaka'wakw, Nisga'a and others.

The museum grounds feature indigenous plants and grasses, as well as two Haida houses and ten full-scale totem poles.

Getting There: The museum is located at 6393 North West Marine Drive on the northwest edge of the University of British Columbia. Between May 21st and September 5th, the museum is open from 10 a.m. to 5 p.m. most days and on Tuesdays to 9 p.m. Winter hours are from 11 a.m. to 5 p.m. Wednesday to Sunday; 11 a.m. to 9 p.m. Tuesdays; closed Mondays. Admission is free year round on Tuesdays from 5–9 p.m.

Stanley Park

THE SHELL MIDDEN covered a huge area—almost two hectares—along the northeast shore of the point; human remains, burial sites and ancient post holes were found by early road builders, and armed forces members arriving in the 1860s were greeted by a village with a large communal lodge and tales of huge potlatches. Despite all this, there has never been an archaeological excavation in Stanley Park, according to the Stanley Park Ecology Society.

It's surprising, for with its enormous cedars and Douglas-firs, its bountiful marshes and lake, its sheltered bays that once drew spawning herring in uncountable numbers, and its fruitful beaches and rocky shores, the point of land that now cradles the park could hardly have been a more perfect place for human habitation.

In fact, what little is known of its history indicates that, until its inhabitants were either killed or driven away by smallpox, there was at least one Musqueam village on the shore of the point near today's Lumbermen's Arch. Known as X'ay'xi (pronounced "Khway-khway"), it was undoubtedly only the most recent of a long line of villages and camps that stretched back into the mists of time. Stumps of ancient cedars more than 500 years old were found

Dennis Fast

growing from one of the middens; the layers of broken shell beneath it were so extensive that the road builders were able to pave their roads with it. This is a place with many stories to tell, if only we care to listen.

But Stanley Park is not only interesting in terms of its human history; it's also geologically fascinating, for it allows us a view of the foundation of downtown Vancouver, as well as a glimpse of what the Fraser Delta might look like far in the future.

The rock along the shores of much of the park is sandstone. Deposited between seventy and forty million years ago by rivers flowing off the newly risen coastal mountains, this is the rock that underlies much of the city centre and can be seen in the facades of its older buildings. In Stanley Park, the sandstone is easily viewed at Ferguson Point, where tidal erosion has given it a stepped appearance, and along the seawall from Third Beach to Prospect Point.

A much photographed exception to all this sandstone is Siwash Rock, at left, or as the Musqueam knew it, *s7xi'lix*, "standing rock", which rises, as though sculpted, from the water on the park's northwest coast. Of hard volcanic rock particularly resistant to erosion, the rock was formed by lava that forced its way through a

Background image: E. Sandys, National Archives of Canada, C-013853

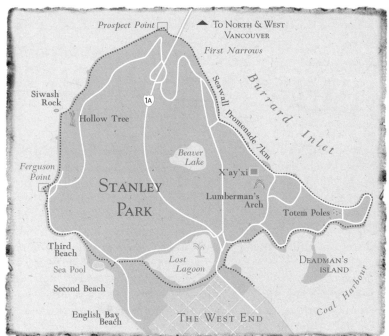

Map labels:
Prospect Point
To North & West Vancouver
First Narrows
Siwash Rock
Hollow Tree
1A
Seawall Promenade 7km
Burrard Inlet
Beaver Lake
Ferguson Point
X'ay'xi
STANLEY PARK
Lumberman's Arch
Totem Poles
Third Beach
Sea Pool
Lost Lagoon
Deadman's Island
Second Beach
Coal Harbour
English Bay Beach
THE WEST END

Allan Taylor

Getting There: Stanley Park can be accessed from north or south via Highway 1A, which goes right through the park. And, though it boasts eight million visitors a year, most during the summer months, it's still possible to get away from it all in many areas of the 400-hectare (or 1,000-acre) park. Many of the trails that wind through the park follow the logging roads created during the park's lumberjack era.

crack or dyke in the sandstone more than thirty million years ago. It once formed a rocky point on the edge of the sandstone, but over time, the wind and waves eroded the softer rock and left the pillar of rock to become one of Vancouver's most recognizable symbols.

There are other, living examples of the park's long history, including the Hollow Tree near Siwash Rock. A "nursery tree", this enormous ancient redcedar is a wonderful example of many of the things that set this species apart from almost all other trees. Sprinkled throughout the park are also culturally modified trees or CMTs (see page 288), as archaeologists term them. Mainly cedars, but also some Douglas-firs and other species, these are trees from which long strips of bark have been taken, to be used for many purposes. The trees have a remarkable ability to heal themselves and, particularly if the bark was taken a century or more ago, those searching for these examples of living history may not, at first glance, immediately recognize them.

The land on which the parks sits is owned by the federal government, which claimed it as a military reserve in 1863, responding to threats—real or imagined—from the Americans who had established a base on the nearby San Juan Islands. The Canadian government leases it to Stanley Park for $1 a year. Though partially logged between the 1860s and 1880s, many trees escaped that fate and for more than a century the point has been an island of green in an ever-larger ocean of urbanity. Though it's unlikely the cougars that once roamed the point's forests will ever return, in recent years, the efforts of the park's ecology society, along with an increasingly "green" provincial population, have together made some impressive strides in reestablishing some of Stanley Park's wild creatures. Bald eagles are once again nesting in the park, and 2004 was the most productive year ever recorded for great blue herons, above, provincially designated a "species at risk".

195

Belcarra Park

BACKED BY THE COAST MOUNTAINS and a climax forest of redcedar, hemlock and fir, and looking west over Indian Arm, with its wealth of coastal and offshore resources, Belcarra Park has all the elements for perennial popularity. At least two dozen nearby archaeological sites, including nine collections of pictographs, attest to the drawing power the region has had for many millennia.

But it was a site at Belcarra Bay, excavated in 1971, that proved to be particularly interesting for an SFU archaeological team led by Arthur Charlton, for it cast light on a little known period in British Columbian history, the centuries between 1,600 and 800 BP. In Europe, much of this time was known as the Dark Ages, and for rather different reasons, those interested in the long view of B.C. history had found themselves similarly in the gloom.

The Belcarra Bay site, located near the picnic area in today's Belcarra Regional Park, was a large midden that was slowly being washed into the sea. Excavating fifteen two-metre-square units down to bedrock,

Charlton and his team found glacial and post-glacial deposits, as well as clear evidence of two distinct periods of occupation, which they called Belcarra I and Belcarra II.

Evidence of the former, which was considerably older, was found in the lower levels of the units closest to the shore. In a layer of charred shell and fire-cracked rock the team recovered a total of fourteen

Working in charcoal in the 1890s, artist E. Sandys captured the misty atmosphere of Indian Arm, a bountiful fjord that inspired an outpouring of art.

E. Sandys, National Archives of Canada, C-013848

leaf shaped and stemmed basalt and quartzite flaked or chipped spear points, and sixteen ground slate points. Flaked choppers and other tools, ground slate knives and adze blades, and hammerstones were also recovered. Only a handful of bone and antler artifacts were found and these were both badly deteriorated and utilitarian in nature—a chisel, a couple of wedges, a possible fish spear and something that might have been a toggling harpoon valve. Based on the design of the points and tools, the older cultural layer was judged to be between 3,200 and 2,200 BP—a period archaeologists call the Locarno Beach phase. It seemed clear that the people who camped here during that long millennium did so to gather shellfish, but not to do much fishing or hunting on the water. And clearly, they came to work and not to party, or even dream of such a thing in the afterlife, for nothing decorative was found with the only burial uncovered.

Belcarra II, dated between 1,650 and 1,000 BP, told a very different story. Here were layer upon layer of sophisticated stone, bone and antler tools and weapons. Among them were beautifully fashioned bone and antler harpoon points, evidence that the people who returned to Belcarra Park repeatedly—likely in the fall and winter—were as much at home on the water as they were on land. The presence of dozens of tiny, side-notched points made it clear that these people—the ancestors of the present Stó:lo Nation—were equally proficient on land and that they had traded their spears in for bows and arrows. But these were not people given to all work and no play, for among the artifacts were beads and pendants, fragments of pipes and a whistle. And beauty was clearly appreciated; even household items such as hand mauls and blanket pins were beautifully finished and sometimes decorated.

The presence of jadeite and other precious stones make it likely that they were traders and the discovery of more than a dozen circular post moulds (the large round posts had long since rotted away) indicated that they may have lived in redcedar longhouses, much as their descendants would a thousand years later.

In short, though the people who inhabited Belcarra Park lived during what many have considered the "Dark Ages", it seems there was nothing "dark" or backward about them. Quite the contrary, they were an accomplished people living in a plenteous region and able to take full advantage of its bounty.

Peter St. John

On the shores of Indian Arm, an enlightened society existed during what Europe called the Dark Ages.

Getting There: From St. Johns Street, Port Moody or Barnet Highway, take the turnoff heading north (Knowles Road). At the second light turn left onto Ioco Road, following the Belcarra Park signs. Turn right onto 1st Avenue and continue as it becomes Bedwell Bay Road. Go past the turnoff to White Pine Beach and turn left at the stop sign on to Tum-tumay-whue-ton Drive. Take the right fork in the road to Belcarra Picnic Area.

197

Allan Taylor

Water creates a gauzy veil as it cascades down the sheer granite wall at Shannon Falls.

Gary Fiegehen

Completed in time for the 2010 Winter Olympics, the magnificent Squamish Lil'wat Cultural Centre in Whistler features the long history of these two peoples, their links to the land and their remarkable art.

The Sea to Sky Highway

Viewed from atop the Stawamus Chief

or from elsewhere along the lower Sea to Sky Highway
Howe Sound is a magnificent testament
to the power of ice and fire.

ONE OF NORTH AMERICA'S southernmost fjords (depending on whether or not you count Burrard Inlet), the sound stretches from Horseshoe Bay on Vancouver's northwestern tip north to Squamish, where it is fed by the Cheakamus and Squamish Rivers. Deep and wide, with mountains rimming its length and islands crowding its entrance to the Strait of Georgia, Howe Sound and its inland valleys together create what Vancouver teacher and author Doreen Armitage described as "a land of superlatives".

Here, thanks to the region's volatile geology, is the world's second-largest granite monolith (see page 202); one of the few places on Earth where one can see the inner core of a volcano; several dormant volcanic cones and mountain slopes that today draw skiers and snowboarders from all over the world.

Here too, thanks to the Cordilleran ice sheet that filled Howe Sound during the last glaciation, are a hanging valley; a sheer wall of lava

500 metres high that holds back an alpine lake; an unusual flat-topped mountain peak; a valley filled with rubble from the collapse of half a volcano and Canada's first underwater marine park.

Most of these features are the result of the eruption of Mount Garibaldi and its associated volcanic cones into and onto the ice sheets that covered the region about 10,000 years ago. Spewing volcanic ash and lava onto the surrounding ice, and then layering further lava onto this hardened base, Garibaldi erected much of its structure on the ice sheet itself. The subsequent melting of the ice (and the concurrent disappearance of the support the glacier had provided for the volcano) led inevitably to the failure of almost half the mountain. Today, the rubble from this catastrophic collapse covers more than twenty-five square kilometres of the Squamish Valley floor. Meanwhile, high above, the inner core of the volcano is exposed. For more information on Garibaldi Provincial Park, see page 205.

The melting of the ice had other consequences. They include a curving undersea ridge about midway up the fjord. This marine moraine marks the furthest advance of the huge tongue of ice that filled the fjord. As the ice melted and froze again in place about 12,800 years ago, sand and gravel spilled off its leading edge, forming a steep permanent hill on the dry,

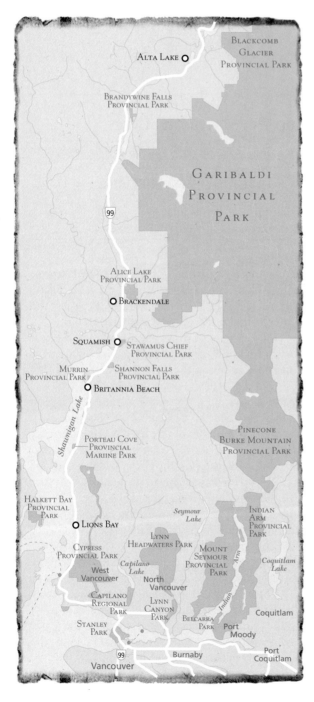

ALTA LAKE ○

BLACKCOMB
GLACIER
PROVINCIAL PARK

BRANDYWINE FALLS
PROVINCIAL PARK

GARIBALDI

PROVINCIAL

PARK

99

ALICE LAKE
PROVINCIAL PARK

○ BRACKENDALE

SQUAMISH ○ STAWAMUS CHIEF
PROVINCIAL PARK

MURRIN
PROVINCIAL PARK

SHANNON FALLS
PROVINCIAL PARK

○ BRITANNIA BEACH

Shaunigan Lake

PORTEAU COVE
PROVINCIAL
MARIINE PARK

PINECONE
BURKE MOUNTAIN
PROVINCIAL PARK

HALKETT BAY
PROVINCIAL
PARK

*Seymour
Lake*

INDIAN
ARM
PROVINCIAL
PARK

○ LIONS BAY

LYNN
HEADWATERS PARK

MOUNT
SEYMOUR
PROVINCIAL
PARK

*Coquitlam
Lake*

CYPRESS
PROVINCIAL PARK

West
Vancouver

*Capilano
Lake*

North
Vancouver

Indian Arm

CAPILANO
REGIONAL
PARK

LYNN
CANYON
PARK

BELCARRA
PARK

Coquitlam

STANLEY
PARK

Port
Moody

99

Burnaby

Port
Coquitlam

Vancouver

cold bottom of the sound. The hill remained when the ice finally withdrew and sea levels rose to reclaim the fjord. As life began to repopulate the sound, undersea creatures—among them corals, sponges, clams, anemones and sea stars of marvellous shapes and colors—were drawn to the shallower, light-filled forests of kelp atop the gravelly undersea ridge.

These in turn drew other marine creatures, including octopi, huge lingcod and shrimp. Before Europeans arrived, the water of the sound roared in the spring with the arrival of millions of spawning herring, drawing minke whales, killer whales, sea lions, fur seals, sea otters and eagles in uncountable numbers. In the summer, pods of migrating humpback whales frolicked in the sound and in the fall, the water seethed with salmon.

This bounty drew human hunters as well; archaeological sites have been found on all the major islands in the sound. Paddlers or hikers can look for pictographs painted on the rock face on the north side of the small bay just past the mouth of Furry Creek, two kilometres north of Porteau Cove.

Today, the waters of the sound are quiet, the vast marine bounty all but extinguished by more than a century of overfishing. Scott Wallace, a diver and researcher at the UBC Fisheries Centre has called it "a silent tragedy". But awareness is building; the conservation area at Porteau Cove Provincial Marine Park and other protected areas are a start at conserving what remain of North America's marine ecosystems, just as national

The Squamish River tumbles toward Howe Sound.

Peter St. John

parks and reserves exist to protect our terrestrial environments.

Other efforts are being made to protect such avian species as bald eagles and peregrine falcons. Residents of Brackendale, a tiny community north of Squamish, boast that more bald eagles call their village home than any other place in the world. The eagles gather by the thousands along the shores of the Squamish, Cheakamus and Mamquam Rivers that surround Brackendale between November and February to feast on spawning salmon. In order to protect these magnificent birds, residents formed the Brackendale Eagles Reserve Society; each year on the first Sunday after New Year's, members gather for the annual eagle count. The one-day total varies each year, but is declining; in 2010, just 956 eagles were counted, less than a quarter of the world record high of 3,766 in 1994.

During the peak period between Christmas and New Year's, it's common to see hundreds of eagles in riverside cottonwoods—often twenty to a tree—or feeding on the sandbars.

Endangered peregrine falcons also nest in the area, including on tiny ledges on the Stawamus Chief. Sections of the huge granite monolith are closed each year to climbers between March 15th and the end of July to protect the nests.

Getting There: The Sea to Sky Highway (No. 99) runs northeast from Horseshoe Bay through Squamish and Whistler to Pemberton and continues as Hwy 99 through the Coast Mountains to Lillooet on the Fraser Canyon. In addition to Porteau Cove, the highway touches several other small provincial parks, including Shannon Falls and Stawamus Chief, which are included in this section, and the huge and magnificent Garibaldi Provincial Park. Others along the way are tiny Murrin Provincial Park, just south of Shannon Falls and Alice Lake Provincial Park just north of Brackendale.

Popular Porteau Cove Provincial Park is located 38 km north of Vancouver on the west side of the highway. The 50-hectare park includes the only public boat launch between Vancouver and Squamish and is a popular site for divers. In the last two decades of the 20th century, several marine vessels were scuttled offshore to provide havens for marine life and enhance underwater exploration. No fishing, shellfish harvesting or removal of other marine life is permitted at the park, which is open all year, with full services from March 1st through October 31st. The name Porteau comes from the French *porte d'eau*, or "water's gate", an apt description when one considers its glacial beginnings. The name dates from 1908 when the region's extensive sand and gravel deposits began to be mined.

The Stawamus Chief

RISING 700 METRES above the town of Squamish, this massive granite cliff is much more than a mecca for rock climbers. It's a peek at what lies below the Earth's surface; this was once molten magma which, when allowed to escape upward through the mantle, cooled into a huge granite batholith. This particular formation, named after the nearby native village of Sta-a-mus, is one of the most recognizable features in what geologists call the Coast Plutonic Complex—one of the largest masses of granite in the world. Indeed some claim it ranks second only to the Rock of Gibralter.

"The Chief", as the locals know it, was created more than ninety million years ago, when the Insular Superterrane (see page 14) slammed into the west coast of North America. Trapped below a ceiling of volcanoes created by the plate tectonics, the magma crystalized. Standing in the parking lot below the cliff face, it's easy to see how the rock was created, for its sweeping vertical formations soar right to the top.

One such dyke, of black basalt, can clearly be seen on the right side of the wall.

But if the Stawamus Chief was born in the fiery heat of the planet's core, it also has other, more frigid stories to tell. At the south end of the parking lot, at the base of the cliff, is a stretch of granite that appears to have been molded and polished. And, in fact, it was, by the huge mass of ice that flowed down Howe Sound during the last glaciation. Moving against the rock face, it scoured and sanded it, in places leaving scratches where large stones scraped against it. The ice changed the Chief in other ways as well, plucking slabs of granite from the north-facing rock and creating the cliff you see today.

Some claim the Stawamus Chief is the world's second-largest granite batholith.

Image released into the public domain. www.wikipedia.com

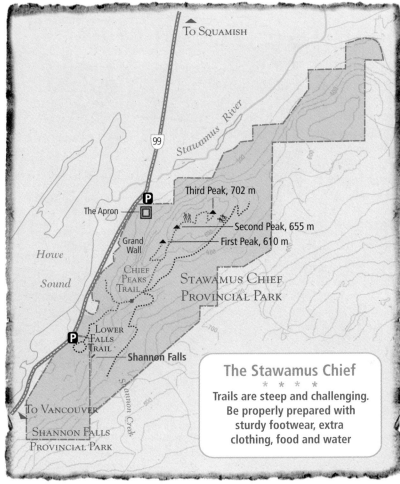

To Squamish

99

Stawamus River

The Apron

Third Peak, 702 m

Second Peak, 655 m

Grand Wall

First Peak, 610 m

Howe

Chief Peaks Trail

Sound

Stawamus Chief Provincial Park

Lower Falls Trail

Shannon Falls

To Vancouver

Shannon Creek

Shannon Falls Provincial Park

The Stawamus Chief
* * * *
Trails are steep and challenging.
Be properly prepared with
sturdy footwear, extra
clothing, food and water

For those who prefer to keep their feet on the ground, there are trails to the summit that range in length from 3.5 to 5.5 kilometres. The shorter route generally takes ninety minutes, the longer one another hour. Both are strenuous, but the view from the top is nothing short of breathtaking.

Getting there:
Stawamus Chief Provincial Park is located just south of Squamish off the Sea to Sky Highway (No. 99), adjacent to Shannon Falls Provincial Park. The forested campground is at the south end of a rough road that hugs the base of

For thousands of climbers and at least one extended family of peregrine falcons, the result is pure perfection. Even if you have no desire to join them, the climbers can often be seen, clinging like ants to the rock face. At last count there were at least 180 routes to the top. But each year from March 15th to the end of July, the climbers avoid one section of the cliff face, in order to allow the endangered falcons, which have nested here since the mid-1980s, a chance to raise their young in peace. So far, the arrangement seems to be working.

the mountain. To view the Chief, take the dirt road up the embankment in the middle of the viewpoint. To reach the trailhead, turn south onto an old north-south road above the viewpoint and continue on to its end.

Hiking from here to the Chief's south summit involves an elevation gain of about 600 metres; you will be climbing almost constantly to the top, but the trail draws upwards of 50,000 hikers a year. If you don't plan to hike, stop at the Stawamus Chief Mountain viewpoint on Hwy 99 in Squamish, a short distance north of Shannon Falls Provincial Park. An interpretive display includes information about the mountain and some of the region's history.

Shannon Falls

A boardwalk trail, inset, leads from the parking lot to the viewing area. In the spring, visitors are likely to be drenched by the spray.

JUST SOUTH of the Stawamus Chief, tiny Shannon Falls Provincial Park surrounds the base of British Columbia's third-highest waterfall. Particularly in spring and fall, the water of Shannon Creek roars over the 335-metre granite wall, drenching those who get too close. Even from the viewing platform, however, visitors can learn a lesson about glaciation from Shannon Falls.

Photos: Allan Taylor

As indicated on page 199, Howe Sound is a fjord that was once steep and narrow, but was greatly broadened and deepened by glacial scouring. At the height of the last glaciation, it was filled with a thick tongue of ice that stretched westward to the widening of the sound near today's Porteau Cove Provincial Marine Park. Branching off this huge glacier were smaller, thinner tributary glaciers, which did not erode as deeply as the main tongue of ice. At Shannon Falls, the result, when the Earth warmed, was a hanging valley high on the wall of the fjord, with a stream that became a picturesque waterfall.

During the summer months, the flow is much less vigorous than it is during the winter and spring, and Shannon Falls is more a lacy veil than a thundering cascade.

From the parking lot off Highway 99, a boardwalk leads to a viewing platform near the base of the falls. Those who would like a closer look can follow a trail that leads from the platform to the base of the falls.

Located three kilometres south of Squamish on the east side of the highway, this is one of the most popular picnic sites along the Sea to Sky Trail, so those wanting to avoid the crowds would be wise to visit in the off-season. From Shannon Falls Provincial Park a one-kilometre trail leads to the base of the Stawamus Chief. To access this trail, park in the lot beside the Logger's Sports Area and look for orange and red markers posted on a large redcedar at the north end of the sports area by the Federation of B.C. Mountain Clubs.

Garibaldi Provincial Park

LIKE MOUNT BAKER, which towers over the southern Vancouver skyline, as well as Mount Meager to the north, Mount Garibaldi is part of the Cascade volcanic arc. At the western edge of a volcanic field that contains at least thirteen vents, Garibaldi has been largely tranquil since the end of the last glaciation. For more than 10,000 years, the region has provided valuable materials—particularly obsidian—for tools, superb hunting grounds, plentiful hot water and magnificent scenery for its human inhabitants. But the mountain's long history has been anything but placid and Garibaldi's fireworks are very likely not over yet.

With origins that go back at least 260,000 years, the story of Garibaldi is one of periods of violent volcanism, followed by long dormant stretches. Beginning about 50,000 years ago, the mountain rebuilt its edifice with a series of fiery eruptions similar to those that issued from Mount Pelée in 1902, destroying the community of St. Pierre on the island of Martinique in the process.

There is clear evidence that Garibaldi's eruptions continued during the last glaciation between about 25,000 and 11,000 years ago. As the ice was beginning to melt, Clinker Peak, a cone on the flank of nearby Mount Price, erupted. Flowing west, the lava met the glacier that filled the Cheakamus River valley and was stopped cold; when the ice in the valley melted, it left a steep, unstable wall of rock almost 250 metres high. Appropriately called The Barrier, the face of the rock wall has repeatedly collapsed over the past 11,000 years, producing large landslides in the aptly dubbed Rubble Creek valley. The most recent slide, in 1855–56, buried a forest and covered an area that Highway 99 crosses today, an obvious reminder to motorists (and highway construction crews) of the danger the mountain continues to pose.

The lava flow also created the basin that cradles Garibaldi Lake, as well as Lesser Garibaldi and Barrier Lakes. In *Vancouver, City on the Edge*, authors John Clague and Bob Turner write that because of the fractured lava "the lakes drain from one to the other by underground

With its clear water and rugged rock, Garabaldi is wilderness at its best.

Allan Taylor

Sure-footed and known for their exceptional speed, mountain goats have been known to climb 450 vertical metres (or 1,500 feet) in less than twenty minutes.

Dennis Fast

channels. At times of high flow, Garibaldi Lake flows into Lesser Garibaldi … by an intermittent stream … [while] Barrier Lake drains underground, and Rubble Creek emerges full-blown from landslide debris at the base of The Barrier."

Farther south, there is evidence that Mount Garibaldi's southwestern flank was built on the glacier that filled the Squamish River valley. When the ice melted, the volcano's western flank collapsed, creating the rugged topography of Atwell Peak and exposing the internal structure of the volcano, along with an enormous debris fan just north of the town of Squamish.

On Garibaldi's southeastern flank is evidence of an even more recent eruption—the Ring Creek lava from Opal Cone. Flowing more than fifteen kilometres down the valley, it left one of the mountain's trademark chutes, with margins marked by raised ridges or levees.

A period of global warming both precipitated and followed the disappearance of the glaciers and

humans may have begun hunting on the mountain slopes as early as 9,000 years ago; several areas above the Squamish and Cheakamus Rivers have long been known as prime habitat for mountain goats. The hunts took place after the rut in late fall, when the animals were in prime condition, with coats that were thick and full. Not surprisingly, the pursuit of these remarkably agile alpine creatures was fraught with danger; only the most experienced hunters attempted it. But success meant not only meat and fat, but status for the hunter. Even more important, the hair was used, often in combination with the wool from specially bred dogs (see page 139) and plant fibres, to make the magnificent and highly valued blankets that were associated with Coast Salish societies for more than 1,500 years. Hunters also pursued deer and elk in the same areas.

Archaeologists have found evidence of hunting parties at two sites not far from Opal Cone, high above the confluence of the two rivers. Both sites are located along what is now known as the Diamond Head Trail, the first on Lava Creek north of Elfin Lakes and the second overlooking Mamquam Lake. Both are small scatters of obsidian or glassy rhyodacite, and seem to indicate that hunting parties stopped to sharpen points or make tools. Situated on ridges overlooking water sources, high alpine meadows and distant ridges and passes, the sites were perfect places to watch for animals—or other hunters—as the members of the party worked.

Obsidian from Garibaldi's volcanoes was highly valued; it has been found in Marpole sites in various places through the Lower Mainland and Gulf of Georgia.

Though there are apparently none within the boundaries of the park, Garibaldi Provincial Park is literally surrounded by hot springs. North and east, along the Lillooet River and its tributaries, and south below Harrison Lake and along the Pitt River are some of B.C.'s finest hot springs. And like the Kootenay Region (see page 60), there is something here for every taste. From the well-developed pools at Harrison Hot Springs Resort to the charmingly primitive Skookumchuck Hot Springs, also called St. Agnes Wells, south of Highway 99, and the remarkable Meager Creek Hot Springs (see page 208), just west of the Lillooet Valley north of No. 99. The largest in B.C.,

in what may be the province's largest geothermal complex, the springs are once again accessible, thanks to a rebuilt road and a new bridge. Like the volcanic field, the cones and lava flows, the hot springs and underlying reservoir are the result of the area's volatile nature and the molten magma that lurks just below.

Though Garibaldi has been quiet for millennia, given its volatile history it's almost certain there will be future eruptions in the region; if so, the accompanying ash columns, lava flows and sudden melting of glacial ice could pose real dangers to air traffic, local communities and wildlife, including the spawning salmon on the Cheakamus, Squamish and Mamquam Rivers.

Getting there: Garibaldi is B.C.'s most popular wilderness park. Both the Diamond Head and Lake Garibaldi trails can be accessed off Hwy 99. The Diamond Head Trail begins at the end of Mamquam Road, which runs 16 km east of the highway at Garibaldi Estates north of Squamish. A lookout before the parking lot provides a panoramic view of the Squamish Valley and Howe Sound. The trail, which attracts hikers, bikers and skiers, leads past Elfin Lakes to north to Opal Cone before turning east to Mamquam Lake. Elfin Lakes features an overnight shelter with 34 beds and a propane and wood stove. A per person, per night fee is charged.

The trail to Lake Garibaldi, known as the Black Tusk Trail, begins at the parking lot off a spur to the east just north of where Hwy 99 crosses Rubble Creek. Constructed well above Rubble Creek, it makes a loop that passes north of the three lakes and connects to spurs to the Black Tusk and other nearby peaks. Two wilderness campgrounds, at Taylor Meadows and the west end of Garibaldi Lake, invite overnight stays. All garbage must be packed out. In winter, ski touring is popular on Lake Garibaldi. A well-maintained trail leads to the Garibaldi Neve Traverse and ultimately to the Diamond Head Trail noted above.

Mount Meager Hot Springs

Fed BY CLEAR, odorless water almost too hot for comfort, the Meager Creek Hot Springs just west of the upper Lillooet River and south of Mount Meager, have undoubtedly been a source of delight for the Lil'wat people of the region for generations. But these springs, along with Pebble Creek Hot Springs farther upstream along the Lillooet River, are also an indication of both the region's volatility and the enormous reservoir of superheated water that lies below.

Mount Meager, the northernmost volcano in the Garibaldi Volcanic Belt, last erupted about 2,350 years ago. But one should not believe that it is dead, for it has a long history of recurring explosions, separated by periods of dormancy. This, along with glacial activity, has made the mountain's slopes highly unstable, something that has been obvious over the past 150 years as five major landslides have occurred on Mount Meager's slopes. Geologists believe even larger ones occurred about 8,700 years ago and 4,400 years ago, removing an enormous section of Pylon Peak and creating a huge natural amphitheatre.

However, Pylon Peak is also the source of an aquifer that may, over the next few years, lead to Canada's first major geothermal power project. The creation of a geothermal

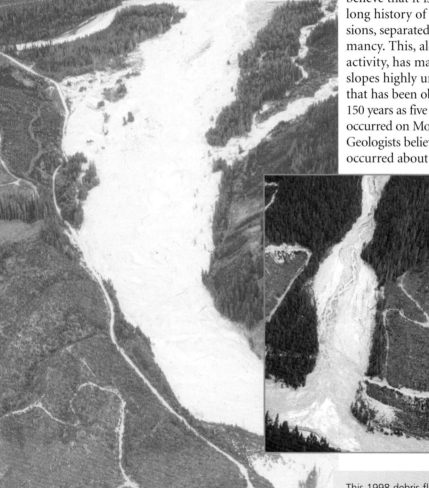

This 1998 debris flow on Mount Meager involved 1.2 million cubic metres of material and dammed Meager Creek below.

Photos: Réjean Couture, Courtesy of Natural Resources Canada

Pebble Creek Hot Springs · Meager Creek Hot Springs · Birkenhead Lake Provincial Park · Birkenhead River · Lillooet River · Place Creek Falls · Pemberton Icefield · Pemberton · Mount Currie · 99 · Nairn Falls Provincial Park · Nairn Falls · Lillooet Lake · Whistler · Garibaldi Provincial Park · Alexandra Falls

Pacific Rim, including the western United States, New Zealand and the Philippines, currently produce geothermal energy for commercial use. Their experiences have demonstrated that, contrary to early expectations, geothermal energy is not always a renewable resource. The key to geothermal use is that the hot water from each aquifer must be extracted at or below the capacity of the underground deposit to replace it.

Despite the construction activity on Pylon Peak just eight kilometres away (and the many workers who might enjoy the popular hot springs at Meager Creek), for a period of time in the early twenty-first century, the springs were not accessible to the public, as a result of an autumn flood in 2003 that washed out the bridge to the Meager Creek springs. The bridge is now rebuilt, and those who visit the springs will find three large cement-bottomed rock pools, a change house and a solar powered composting toilet.

The springs have been designated for day use only—8 a.m. to 8 p.m.

aquifer of this kind requires very specific conditions. They include older bedrock to act as a basin for the water; a significant water source to fill and refill the basin; magma that rises close to the surface to heat the surrounding rocks, and a caprock—in this case a volcanic intrusion inside the mountain, to keep most of it from escaping. The water that does escape along fault lines in the bed-rock rises to the surface to create hot springs such as those at Meager Creek.

All these conditions are apparently present on a steep slope below Pylon Peak. Test wells drilled by BC Hydro in 2003 between No Good Creek and Angel Creek indicated the presence of a large "high temperature" reservoir; water temperatures here were up to 275 °C. The following year, believing the site has an energy potential of at least 200 megawatts—enough energy to service at least 80,000 households—the government initiated two production wells.

Other countries, particularly those around volatile edges of the

Getting there: Mount Meager is located about 150 kilometres north of Vancouver, west of the Upper Lillooet Forest Service Road, which diverges north of the Pemberton Valley road at the 23-km mark. The Meager Creek Hot Springs are accessed by a branch road at the 39-km mark. Watch for the kilometre signs on the road. Turn left, drive over the Lillooet River and continue for a further 8 km, staying to the left at the fork. Park in the parking lot (not near the new bridge over Meager Creek. Once across the bridge, take the trail on the right to the springs. Clothing is optional, but dogs are not allowed, since authorities have threatened to close the springs for health reasons if dogs are permitted.

The Keyhole (or Pebble Creek) Hot Springs are 5.4 km farther along the forestry road, at the end of a long flat stretch, just before the road climbs away from the river. From here, you walk along the river about 1.5 km to where several hot water vents flow from the bank above the river. You must wait until after spring runoff to access these springs.

THE
GULF
ISLANDS

Steaming through Active Pass, a BC Ferries ship is
bound for Vancouver.

Peter St. John

The sandstone bedrock of the Gulf Islands can be clearly seen at Montague Harbour on Galiano Island.

The Gulf Islands

Island hopping

by ferry, water taxi, float plane or kayak
to the ever more popular southern Gulf Islands
offers abundant opportunities for close encounters.

THE ISLANDS MAKE it easy to be close up and personal with some of the some of the West Coast's most interesting geology, paleontology and archaeology. But while digging in their fascinating past, don't forget to make time for the islands' striking beauty. Or their eclectic coffee houses, art galleries and bookstores.

Geologically speaking, the southern Gulf Islands are among B.C.'s latest acquisitions, having risen from the depths of the Georgia Basin about forty million years ago. But farther north, as the Hornby Island section (see page 222) by Heidi Henderson and Phillip Torrens clearly illustrates, the fossils of fascinating creatures from much earlier eras can be found here as well.

The Gulf Islands also continue to be of interest to archaeologists. Not only do many of them, including Gabriola, Pender and Galiano, have obvious evidence of seasonal or year-round occupation by Coast Salish peoples as far as back 5,000 years,

underwater archaeological excavations in Montague Harbour on Galiano (see page 215), have proven that shellfish were regularly harvested much earlier. And occasional discoveries of large spear points make it likely that hunters were quietly stalking deer and elk on some of the larger islands even before 7,500 BP.

On several islands, particularly Gabriola (see page 220), magnificent petroglyphs make it clear that these were places of special significance to the Sneneymuxw and other Hul'qumi'num people—"those who speak the same language". Though rock carvings are notoriously difficult to date, many on Gabriola were deeply covered with moss or lichens when they were rediscovered and some were so worn that the images could barely be seen; both are indications of considerable age. To save the original carvings, and yet share these images from the past, the Gabriola Museum has undertaken the task of reproducing many of the petroglyphs and, with the blessing of the Snuneymuxw people who created them, has made them accessible to visitors to the museum grounds. For those wishing to copy the reproductions for personal use and pleasure, rubbing kits are available at the museum.

On Saltspring Island, treasured antiquities of another kind are being protected. Here, some of the last significant tracts of B.C.'s most endangered

Bald eagles are unique to North America and Canada's largest birds of prey. The nation's largest breeding population is found in British Columbia.

Dennis Fast

ecosystems, as well as red- and blue-listed plants and animals have recently, perhaps just in time, been protected from development with the creation of linked parklands around Burgoyne Bay. Ultimately, the goal of islanders is to create a park that stretches from "mountain to mountain and sea to sea", encompassing

SOUTHERN GULF ISLANDS

THETIS ISLAND

KUPER ISLAND

GALIANO ISLAND

FERRY TO TSAWWESSEN

CHEMAINUS

VESUVIUS

LONG HARBOUR

VANCOUVER ISLAND

GANGES

MAYNE ISLAND

SALTSPRING ISLAND

NORTH PENDER ISLAND

SATURNA ISLAND

DUNCAN

FULFORD HARBOUR

SOUTH PENDER ISLAND

COWICHAN BAY

FERRY ROUTES

the Gulf Islands' two highest peaks and stretching from Sansum Narrows on the west coast of the island to Captain Passage on the east.

Most of the islands are easily accessed by ferry, but visitors bringing cars must be prepared to slow down; not only do the islands' sites deserve it, their roads demand it. Narrow, winding and often hilly, the only roads on many islands are also someone's front street. Children may be playing, walkers walking or bikers biking and speed limits are generally what one would find in urban residential neighborhoods. Information on each island is easy to find, but don't hesitate to ask directions; islanders are proud of their homes and are usually happy to assist visitors in discovering why.

214

Galiano Island

LIKE GABRIOLA TO THE NORTH, Pender to the south and Saltspring (or Salt Spring, as the locals prefer) just to the west, the bedrock of Galiano is largely sandstone, conglomerate and shale, sediments that were deposited just offshore of an ancient delta following the "docking" of Wrangellia about ninety million years ago.

And like its neighbors, Galiano lies on a northwest-southeast line, with ridges of sandstone and valleys of shale. The reason for this alignment has to do with the impact of wave energy in the Strait of Georgia on the bedrock of the islands. Because sandstone (which is composed of fine grains of feldspar or quartz that are cemented with even finer clay under pressure and over time) is much harder and more resistant to weathering than shale, the wave action has eroded the shale, leaving long, linear outcroppings of sandstone.

In many places on Galiano, these ridges of sandstone run right into and under the surrounding waters of the strait. One of these, Montague Harbour along the southeast coast of the island, has clearly drawn people for thousands of years. Even today, the reasons are clear; with its tidal lagoons, open meadows and towering forests, the area is a haven for plants and birds. And though harvesting shellfish is no longer allowed due to the poor water quality within the harbor, for thousands of years this was clearly not the case; just south of the parking lot, an eroded shell midden can be seen all along the water's edge. And otters can often be seen fishing in the bay.

Early archaeological excavations into this large deposit of broken shell seemed to indicate that the midden dated back perhaps 3,500 or 4,000 years. However, even back in the 1960s, archaeologists also noted that the midden was clearly eroding into the sea and that the oldest deposits extended well beyond the high water mark. For decades, the questions remained: was Montague Harbour not inhabited prior to what is referred to as the Locarno Beach period? Or were the deposits from earlier habitations now under water?

It took Norm Easton, an anthropologist from Yukon College in Whitehorse, to provide the initiative to find out. Beginning with a series of small corings in 1989, Easton discovered that there were indeed older deposits off shore. Over the succeeding years, with federal and provincial assistance, he

MONTAGUE HARBOUR PROVINCIAL MARINE PARK

Campground

Lagoon

Gray Peninsula

Montague Harbour

Porlier Pass Road

90 m

Archaeological Dig

Getting there: With its hills and harbors, bluffs and beaches, and both terrestrial and marine provincial parks, Galiano is a popular spot to visit. From the heights of Mt. Galiano, ferries from Vancouver and Vancouver Island can be seen passing through Active Pass to the Sturdies Bay ferry port at the north end of the pass. Galiano can also be reached from Mayne, Pender and Saltspring Islands by water taxi. To visit Montague Harbour, and its fine provincial marine park, take Sturdies Bay Road west to Porlier Pass Road and follow it southwest to Montague Road. Turn left or east and follow the signs to the park. The park includes trails and a magnificent rock ledge on the northwest edge of Gray Peninsula, which has been carved by glacial movement into rippling patterns. Continuing on Porlier Pass Road leads to the north end of the island, Porlier Pass and the Penelakut Reserve. Dionisio Point Provincial Park at the island's northeast corner has sea access only. To get to Georgeson Bay, turn off Sturdies Bay Road onto Georgeson Bay road and follow it south. Galiano is hilly and forested, with winding, narrow roads and restricted visibility. The speed limit is 50 kph all over the island and cyclists are warned to use caution.

215

Cedars are reflected in the still water of Montague Harbour.

Peter St. John

directed the first underwater archaeological excavations in British Columbia. The results were quite stunning. They showed that with sea levels between three and ten metres lower during the mid-Holocene (from 8,000 to 5,000 thousand years ago), campsites established to harvest shellfish were as much as ninety metres off today's shoreline.

Using a caisson—a watertight box that could be pumped dry of water to allow the excavation to be carried out—the team discovered spear points, fish hooks, bone beads, scrapers and charcoal dating back more than 6,500 years. Just as important, Easton's initiative not only led to other underwater archaeological excavations—in Haida Gwaii, among other places—but contributed to a basic rethinking of the peopling of North America and a new appreciation of the idea of a coastal migration from Asia (see page 22).

The early inhabitants of Galiano Island were not only gatherers of shellfish, however. They were also very much at home on the surrounding waters, including Porlier Pass, which separates Galiano from Valdes Island to the northwest. Like Race Rocks in the Juan de Fuca Strait (see page 251), Porlier Pass is a high-current area, with shallow reefs and outcropping rocks. During flood tides, the current here reaches speeds of nine knots, which delivers a steady supply of nutrients to a great abundance of marine life, just as it does at Race Rocks. Salmon, lingcod, rockfish, flatfish and herring all frequent the pass or make it their home. And, according to UBC librarian, avocational archaeologist and keen scuba diver Erik de Bruijn, as a result of the abundant fish, "[a]t certain times of the year, marine mammals, including seals, sea lions and even killer whales frequent the waters of the pass."

Writing in *The Midden*, the publication of the Archaeological Society of British Columbia, de Bruijn provided rock hard evidence that the inhabitants of Galiano have been fishing the bountiful waters of the pass for centuries, or even millennia. Off Alcala Point (which is on Penelakut reserve land) he found a perforated anchor stone on a sloping shelf of broken rock in eleven metres of water. Weighing a hefty 10.8 kilos, triangular in shape, with a hole for a cord painstakingly pecked through the apex, it was so encrusted with marine growth that the hole could hardly be seen. Though it could not be dated, a similar sandstone weight found at Montague Harbour was dated to about 2,000 BP.

De Bruijn also found a T-grooved stone weight near Black Rock, just off the tip of Valdes Island in Porlier Pass. Though no other weights were found in the area, four similar T-grooved weights were excavated at Georgeson Bay at the south end of Galiano, in an archaeological component that was dated to between 3,100 and 2,820 BP. Anchors of various designs very likely either held down lines of halibut hooks or anchored nets to prevent fish traps from moving.

Fishers appear to have come on the heels of even earlier hunters. At Dionisio Point, just northeast of Porlier Pass, excavations unveiled a spear point, which was dated to between 9,000 and 7,500 BP.

PENDER ISLAND PORTAGE
Near this point passed an ancient
trail over which Indians portaged
their canoes between Browning and
...... harbours. Across this neck
...... settlers later
dragged their boats on skids for
visits between the scattered island
families or to shorten the journey

Pender Island

TODAY THERE IS A TINY BRIDGE spanning a man-made canal between the north and south halves of Pender Island, but in the past a neck of land—Xelisen (pronounced Hel-isen), "lying between"—connected the two parts of the island. On either side of this isthmus lay two of the most

Photos: Peter St. John

bountiful shellfish harvesting sites in the Gulf Islands.

The proof of this plenty can be easily seen from the beaches on the northwest and southwest sides of the isthmus. On the northwest, just metres from the waters of Browning Harbour, a veritable cliff of discarded shells rises from the shingled shore.

Behind it, the midden continues upward in a mound. On the southwest, from Canal Road to Bedwell Harbour, is another mounded midden that has proven to be even older. Yet these are only remnants of the huge mounds of shells that existed before the canal was dredged through the isthmus in 1903. Since then, storms and tides have combined to eat away at what remained. It was this rapid erosion of one of the most obviously ancient sites on the islands that led Simon Fraser University archaeologist Roy Carlson to recommend excavation of the site in the early 1980s.

Over three summers between 1984 and 1986, in a project jointly funded by the province and the university, excavations directed by Carlson and Philip Hobler revealed that Xelisen—or the Pender Canal Site, as the archaeologists called it—had been used as a spring and summer encampment for harvesting the resources of the sea for more than 5,000 years.

Among the many artifacts recovered from the excavations were labrets—lip ornaments, rather like many young people wear today, though made of stone rather than silver or gold—dating back almost 4,500 years, as well as net and line sinkers, fishing and sea mammal hunting tools, adzes and hammers for woodworking, ornaments of bone, shell, soapstone and—discovered on the final day of the dig in 1985—four beautifully carved spoons made of antler.

The older of the two huge middens began just south of today's Canal Road, and stretched almost to the beach on Bedwell Harbour. Here, in addition to one date in excess of 5,000 BP, were a large group of artifacts dating to between 4,430 and 2,580 BP and another significant collection that proved to be between 1,400 and 1,100 years old. Within this section of the midden, as well as the other, just north of today's Canal Road, the archaeologists also found more than 100 burials, including sixty-seven men and

217

John Harvey

Hunted for pelts and bounty, harbor seal populations were threatened by 1970. Today, their numbers have recovered and seals can once again be seen in the Gulf Islands.

women whose remains were complete enough to study. Four of these were radiocarbon dated to between 5,000 and 4,000 BP.

Remarkably, the majority were over the age of forty-five when they died and many, subsequent skeletal investigation by osteologist Erin Strutt found, had recovered from fractures to body parts literally from head to toe. By comparison, the average life span of Europeans of the same periods ranged between just eighteen years during the Bronze Age (about 4,000 to 2,500 BP) and thirty-three years of age in Medieval Britain. It's also interesting to note that none of the identifiable bones excavated in the Pender Canal Site were those of children. This might be interpreted in a number of ways, but clearly, the absence of communicable diseases, as well as healthy diets based on fish and sea food, made a huge difference to the survival rates and average life span in pre-contact British Columbia.

The shell middens were also littered with the bones of both land and sea mammals, allowing archaeologist Diane Hanson to create a picture of the prey that was hunted or gathered by the Pender Islanders. Unlike many of their mainland contemporaries, salmon was of little importance in the diet of the islanders, at least during the spring and summer months. Instead, in late winter and early spring they fished for herring, and also for the fish and sea mammals that preyed on the herring, including cod and harbor seals. They also fished for sea perch and rockfish as well as other species. On land, they hunted deer and used every part of

the carcass. But mainly, Pender Island was a place for gathering shellfish in enormous quantities. These included butter clams and blue mussels, as well as limpets, dogwinkles, urchins and rock crabs. These were eaten fresh, smoked or dried.

Getting there: With dozens of accessible beaches and sheltered coves as well as several large woodlands that are part of the Gulf Islands National Park Reserve, Pender Island is a popular vacation spot. It can be easily reached by ferry from Tsawwassen on the Lower Mainland or Swartz Bay on Vancouver Island, as well as from Long Harbour on Saltspring Island. Float plane services are also available from Vancouver, Victoria and Seattle (U.S. Customs is located at Bedwell Harbour.) To reach the Pender Canal Site from the ferry terminal, follow Otter Bay road east to Bedwell Harbour Road, then turn south through Driftwood Centre (the main shopping centre). Soon after, the road becomes Canal Road, which turns sharply east just before the bridge. A plaque with information about Xelisen can be found on the north side of the road, just west of the bridge.

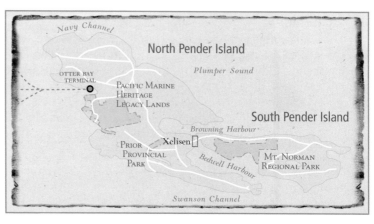

Saving Saltspring's Special Places

SAVING SALTSPRING

BURGOYNE BAY, which indents Saltspring Island's southwest coast from Sansum Narrows, is the largest remaining undeveloped area in the ever more popular southern Gulf Islands. And towering (at least in Strait of Georgia terms) over it, are two of the highest points in the region—Mount Bruce, at 709 metres, to the south and Mount Maxwell, at 594 metres, to the north.

On the slopes of Mount Maxwell are more than thirty per cent of the Gulf Islands' remaining Garry oak woodlands, along with a small grove of endangered old-growth coastal Douglas-fir. The Garry oak ecosystem is one of Canada's four most endangered environments (see page 241), while only one per cent of old-growth Douglas-fir habitat remains intact in B.C. Buffering this precious old-growth forest is a large expanse of second-growth Douglas-fir. Together, they provide habitat for thirteen endangered plant and animal species.

The waters of the bay are no less precious; with healthy tidal flats and eelgrass beds, they draw orcas, harbor porpoises and seals. Dozens of species of birds, including western grebes, bald eagles, peregrine falcons and great blue herons all nest nearby and two salmon streams run into the bay, a rarity in the Gulf Islands.

Somehow, though Saltspring is the most populous of the southern Gulf Islands with more than 10,000 residents, Burgoyne Bay and its sur-

Getting there: With the main ferry terminal at Fulford just five nautical miles from Swartz Bay on Vancouver Island, sailing time to Saltspring is about a half-hour. There are eight sailings a day, in both directions. To reach Burgoyne Bay from the ferry terminal, take the Fulford-Ganges Road northwest to Burgoyne Bay Road and follow it west to the bay. Saltspring can also be reached by ferry from Crofton, south of Chemainus. Sailings are almost hourly.

rounds have remained largely untouched. When Vancouver developers purchased a significant proportion of the land around the bay in 1999, a number of organizations and government departments realized, just in time, that the region was nothing short of priceless. Since then, using provincial and federal money, along with funds provided by the Land Conservancy of British Columbia, 770 hectares of land on the bay's north and south shores were purchased. With existing provincial and regional parks, the parkland in the area now totals more than 10,000 hectares. In fact, slightly more than a quarter of the land on Saltspring is considered to be in its natural state and about ten per cent is fully protected.

Ideally, the island's residents (many of whom prefer the old spelling—Salt Spring) would like to see the Burgoyne Bay region added to the new Gulf Islands National Park, which was established in 2003. To date, that hasn't happened, but islanders are optimistic. They know what they have is worth saving for the future.

Gabriola Island

LIKE MUCH OF BRITISH COLUMBIA, Gabriola began life underwater. But while some of its near neighbors, including Vancouver Island, were born as volcanic island arcs (see page 10) far out in the Pacific, Gabriola (and other nearby islands including Valdes and Galiano) began as part of an enormous submarine delta. The delta sediments were deposited at the end of the Cretaceous—between sixty-five and seventy-five million years ago—by a great river that flowed from the ancient North American continent.

But let's backtrack for a moment. Some twenty-five million years before, about ninety million years ago, Wrangellia—the largest part of an island arc or terrane known as the Insular Superterrane—"docked" on the western edge of Laurentia. At the time, North America was somewhat farther south and the climate was considerably balmier than it is today (even on the Gulf Islands). The arrival of Wrangellia, which included Vancouver Island and the accompanying sea floor, caused considerable upheaval inland, but over millions of years, it securely attached itself to the continental margin. This permanent link to North America was below sea level and not only created the Georgia basin, a wide enclosed gulf between Vancouver Island and the mainand, but forced the subduction zone—the meeting place between the Pacific plate and the continental plate—west of Vancouver Island, where it remains today.

It was into the Georgia Basin that the sediments of the submarine fan were deposited. And over time, they created an enormous delta, rather like the one that is rapidly growing today at the mouth of the Fraser River or, according to Nick Doe, editor of *Shale*, the excellent journal of the Gabriola Historical and Museum Society, perhaps more akin to the "seven deltas [that have been] used by the Mississippi River at various times in the last six thousand years". As the millennia passed, the submarine fan grew. Eventually, it was nearly five kilometres thick, comparable, wrote Doe in the January 2004 issue of his journal, to today's Amazon River delta.

Reflecting the way such deltas are built, and resulting from the mountain building and subsequent erosion that were going on inland on the North American continent, in some places or over some periods, the submarine fan was comprised of fine silt or sand and elsewhere was made up of course sand, pebbles and rocks. As the sediments flowed down the river into the sea, they buried earlier layers, compressing them and ultimately turning them to mudstone or shale, sandstone and conglomerate. For millions of years all this lay at the bottom of a wide gulf.

Then, in a pair of collisions fifty-four and forty-two million years ago, two smaller land masses, the Pacific Rim Terrane and the Crescent Terrane, collided with the southwest corner of Vancouver Island (see page 258).

Peter St. John

220

An ocher starfish awaits the coming tide. As befits the name, these beautiful echinoderms come in purple, brown and yellow.

These collisions had far-reaching consequences; they added some of the most scenic parts of Vancouver Island and, through volcanic action, they created the "plumbing" for a mini-gold boom in the island's southwest corner. More important, as far as Gulf Islanders are concerned, they pushed Vancouver Island closer to the mainland and, in the process folded and tilted the sediments in the Georgia basin until they poked above the waters of the gulf. *Voilà!* The Gulf Islands.

During the last two million years, Gabriola's marine sandstone (which is named for the island) has been eroded, planed and smoothed by successive glaciations, until today the island, when viewed from the sea, resembles a shallow platter. Below the thick upper layer of Gabriola Formation sandstone is the mudrock or shale of the Spray Formation, which weathers more easily. Layered outcroppings of this formation can be seen along Leboeuf Bay, on the northeast shore of the island. Spray Formation shale can also be found on Hornby Island (see page 222); it is, in fact, named for Spray Point on Hornby's Tribune Bay.

Sandstone of the Gabriola Formation also erodes, though less from active weathering than as a result of crystallizing salt water, which causes the clay holding the tiny particles of sand together to evaporate. The end-

product in many places is honeycombed sandstone that looks like lacework. The best known example of this ornamental sandstone is the Malaspina Galleries on the northwest corner of the island.

Sandstone has also played a significant role in the island's human history, for it provided a perfect slate (pardon the pun) for Gabriola's many magnificent petroglyphs. More than seventy examples of this rock art have been uncovered on Gabriola, earning it the nickname "Petroglyph Island". Because the many beautiful carvings have a special, even sacred significance for the Sneneymux people, the original inhabitants of the island, and because they can easily be eroded when exposed to weathering and the creation of "rubbings" or destroyed by vandalism, the Gabriola Historical and Museum Society began in the 1990s to make reproductions of the petroglyphs. Today, several dozen can be seen on the grounds of the museum, where visitors are invited to view them, learn more about their history and take photographs.

Getting there: Gabriola can be easily reached by ferry from Nanaimo. Ferries depart almost hourly, with a crossing time of 20 minutes. The Gabriola Museum is just minutes from the ferry dock on South Road, while Malaspina Galleries can be reached by taking Malaspina Drive off Taylor Bay Road, just east of the ferry dock. A trail leads from the end of the drive down onto Malaspina Point. However, exploration of the seaside formations was prohibited in 2004, due to concern that the overhanging formations were increasingly unstable. At low tide, visitors can still stand on the shore at one end and look into the cavern, but the best way to view the formations is now by kayak. Taylor Bay Road passes through Descanso Regional Park, with its campground, and leads to Gabriola Sands Provincial Park.

Photos: Peter St. John

Gabriola Formation sandstone falls in lovely layers to the sea at Drumbeg Provincial Park.

Hornby Island by Philip Torrens and Heidi Henderson

HORNBY ISLAND, one of the northern Gulf Islands nestled between Vancouver Island and the British Columbian mainland, is a mecca for tourists, folk musicians, and scuba divers alike. The divers are drawn by the rare opportunity to see six gill sharks that have come up into shallow waters. In most times and places, *Hexanchus griseus*, as six gills are formally known, live thousands of feet beneath the surface. For a few months a year, for reasons not fully understood, they venture up into waters less than thirty-five metres deep off Hornby.

But you don't have to don scuba gear to explore B.C.'s rich marine life, present or past. Killer whales can be seen frolicking off the coast during their summer migration. Twice a day the sea recedes, offering up tidal pools teeming with flora and fauna, exposing colorful kelp, crabs and starfish, all busily making their living in these nutrient-rich waters. The beaches are also home to flocks of seagulls, which fall like airborne armies onto exposed shellfish, starfish, and dead fish. With a climate rather warmer and drier than nearby Vancouver Island, many eagles and herons choose to live here year-round. A variety of seabirds abound in fall, winter and spring with nearby Metcalf Bay, on Denman Island, acting as a wintering ground for thousands of loons, grebes, gulls, cormorants and ducks.

These modern seabirds bring to mind pterosaurs, which scientists believe occupied similar ecological niches. Like birds, pterosaurs (popularly known as pterodactyls) came in a wide variety of sizes, including some with wingspans of more than six metres—making them the largest living things that ever flew. They ruled the Jurassic skies some 200 million years ago.

Because they evolved from reptiles prior to modern birds, it was once believed that pterosaurs were primitive, passive fliers. They were imagined as gliding, rather than actively beating their wings. We now know that they were powerful fliers, chasing and catching their prey on the wing. One clue to this revelation is a small bone at the front of the wing bone that curves back toward the shoulder, rather like an elongated thumb on a spread hand. Modern birds have a small but vital feather, the aula, in this position. It shifts, acting like the leading edge on some airplane wings, redirecting the airflow over the wing and, with comparatively little effort, allowing changes in speed and angle in the air. It seems the pterosaurs' extended thumb would have held a flap of membrane in a similar position at the front of the wing, for a similar purpose.

The other clue that pterosaurs were sophisticated flyers comes from their skulls; they have much larger brain cases in relation to their size than their earth-bound contemporaries. Coordination of flight requires tremendous brainpower, and coordination of active flight, with constant shifting in the shape and location of massive wings, even more so. Nature is extremely parsimonious, not

Fossil photos Dennis Fast

Jack Most

frittering away investment in any organ where it is not needed. Given the engineering challenges and the energy costs of getting each additional gram of weight off the ground, it's unlikely that pterosaurs would have developed such large and heavy "on board computers" unless they clearly paid their own way in faster, more nimble flight that would have allowed their owners to catch more prey and outmaneuver competing aerial hunters and scavengers.

Imagine flocks of pterosaurs weaving and wheeling as they jockeyed for the lead and the feast that accompanied it or fought to be the first to snatch up the low tide offerings.

Today, amateur and professional paleontologists alike avail themselves of similar tidal opportunities. The rise and fall of the sea wears away the soft shale, regularly revealing new buried treasures, like a merchant

Among British Columbia's best-loved animals and best-known symbols, orcas can still be seen in the Strait of Georgia off Hornby Island. But their numbers are in decline and they have been officially designated as members of an endangered species.

constantly restocking the shelves with tempting new wares. Here one can find fossils of clams, snails and shark's teeth, beneath the apron of shale that forms the beach. Many fossils are also found inside exposed concretions—small, rounded calcarious nodules that rest in the intertidal zone.

While the sea swirls, rising and falling twice daily, the larger tectonic plates do their own, much slower dance to the music of time, shifting, uplifting and unearthing large chunks of the sea floor. The rocks of Hornby were early steps in this dance, laid down just after the volcanic range that would become Vancouver Island and Haida Gwaii collided with what would become B.C. From dating volcanic ash within the deposits, we know they were formed sixty-seven million years ago, hundreds of metres beneath the Pacific.

THE SIX GILL— *A Jurassic Shark*

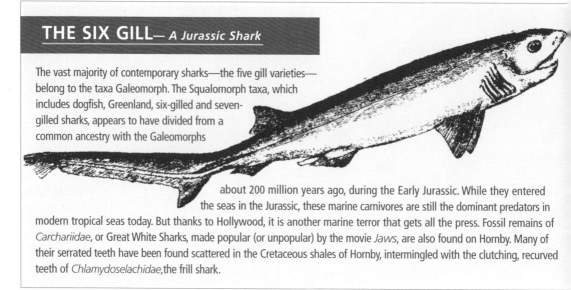

The vast majority of contemporary sharks—the five gill varieties—belong to the taxa Galeomorph. The Squalomorph taxa, which includes dogfish, Greenland, six-gilled and seven-gilled sharks, appears to have divided from a common ancestry with the Galeomorphs about 200 million years ago, during the Early Jurassic. While they entered the seas in the Jurassic, these marine carnivores are still the dominant predators in modern tropical seas today. But thanks to Hollywood, it is another marine terror that gets all the press. Fossil remains of *Carchariidae*, or Great White Sharks, made popular (or unpopular) by the movie *Jaws*, are also found on Hornby. Many of their serrated teeth have been found scattered in the Cretaceous shales of Hornby, intermingled with the clutching, recurved teeth of *Chlamydoselachidae*, the frill shark.

Getting There: To get to Hornby Island, head north on the Island Highway to Buckley Bay, 20 km south of Courtenay. BC Ferries operates a scheduled service from

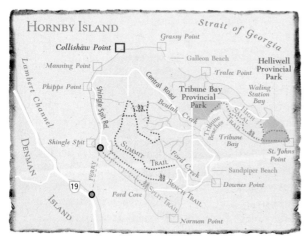

Buckley Bay to Denman Island. Travellers wishing to visit Hornby Island must first take the ferry to Denman Island and drive across the island to Gravelly Bay on the east coast of Denman Island. A second ferry to Hornby Island makes the short crossing between Gravelly Bay on Denman and Shingle Spit on Hornby Island.

Finding fossils and petroglyphs on Hornby Island: To reach the popular fossil collecting site at Collishaw Point from the ferry terminal, follow Shingle Spit Road through a sharp turn as it becomes Central Road. After 500 metres turn left onto Savoie Road and drive north and park at the turnaround. A trail to the beach follows a public access on the left side of the fence. On reaching the beach, turn right to face Collishaw Point, known locally as Boulder Point. Look for bits of fossil shell material in the shales at your feet and also in the plentiful concretions littering the beach. If you find vertebrate material or something unusual, bring it to one of the local Vancouver Island museums for inspection. You could have a new species named after you. To view the Hornby petroglyphs, take the road to Tralee Point at a good, low tide.

Petroglyphs by Philip Torrens and Heidi Henderson

CHE'TIYCHIN, as the Sliammon First Nations peoples knew Hornby Island, is home to several petroglyphs. In contrast to pictographs, which are paintings on rock, petroglyphs are made by carving designs into the rock, using stone tools. Petroglyphs are enigmatic in both age and meaning.

Because petroglyphs are carved from rock rather than from bone or other organic matter, archaeologists cannot determine their ages using carbon-14 or other radioactive dating techniques. Attempts to determine ages by erosion are challenged by the fact that we have no way of knowing how deeply carved the grooves were in the first place. Given the heavy rain on most of the B.C. coast, it seems improbable that any surviving petroglyphs date back to the beginning of human presence here—at least 12,000 years ago. Estimates of their ages range from a few thousand years to less than a century, depending on the petroglyph and its location.

Petroglyphs are equally cryptic in their purpose. They are not believed to be writing, or even necessarily to be literal representations. As Grant Keddie, curator of archaeology at the Royal British Columbia Museum pointed out in a 1998 article in *Discovery* magazine, it is vital to bear in mind that for First Nations peoples the natural and supernatural worlds were inextricably intertwined, not separate, as European cultures tend to view them. The supernatural world could affect events in this world, just as human actions could affect the supernatural world—and thereby, this natural world, as causes came full circle. Thus, many petroglyphs may have had ritual or magical significance. Others, like European statues, appear to have commemorated momentous events.

Photos: © Marja-Leena Rathje / www.marja-leena rathje.info

Petroglyphs can be created by pecking or grooving the rock into shapes, or by chiselling away the surrounding surface to reveal an image. Both methods are enormously time consuming, but the result lasts for centuries … or longer.

VANCOUVER ISLAND

A visitor strolls among climax Douglas-firs at Cathedral Grove.

Photos: Dennis Fast

With its brimming baskets and ornate Legislative Building, this is a view people commonly associate with Victoria. But there is much more to B.C.'s capital, for it has a long and storied past.

Victoria

Residents of Greater Victoria

pride themselves on their genteel surroundings,

magnificent scenery, balmy winters

and charming architecture.

AND CERTAINLY, it's hard to argue. But beneath all this refinement, and fundamental to the region's striking panoramas, is nothing less than geological mayhem.

Like most of the rest of British Columbia, southern Vancouver Island and the picturesque islands and waterways that surround it were created by continental collisions, which produced violent earthquakes and explosive volcanoes. This cataclysmic creation process went on for more than sixty million years and indeed, even today the earthquakes continue at an average of 200 a year.

The vast majority are too weak to be felt; only a handful are strong enough to be noticed. Those large enough to cause substantial damage occur perhaps once a decade, while earthquakes measuring 8 or more on the Richter Scale generally happen centuries apart.

Vancouver Island was once one of a series of volcanic terranes, or island arcs, situated far to the south-

west in the Pacific Ocean. Picture a primeval version of today's Indonesia. Sliding northeast on the Pacific Plate, the largest of these archipelagos—Wrangellia—collided about ninety million years ago with the already enlarged edge of ancient North America, which was moving slowly southwestward. The force of this colossal collision drove the edge of Wrangellia downward, creating the Georgia Depression, today's Strait of Georgia and Puget Sound. Turning to molten lava beneath the edge of the continent, the subterranean coast of Wrangellia then created a chain of volcanoes where the Coast and Cassiar Mountains are today. Much farther inland, the impact created front ranges of the Rocky Mountains, which had been born when larger, earlier terranes collided with the ancient continent of North America.

About fifty million years after Wrangellia thus "docked" with the North American coast, another, smaller land mass, the Pacific Rim Terrane, collided with Wrangellia's southwest coast. The fault line where the two islands met can still be seen today in the San Juan and Survey Mountain Faults, which run from Port Renfrew to beyond Cobble Hill, and down in an arc to Bamberton.

Around the same time, volcanic activity on the seabed to the west of Vancouver Island gave birth to the Crescent Terrane. Over time, the volcanoes crested the waves and grew into a tiny volcanic island about the size of Iceland and a series of fledgling sea floor volcanoes. About forty-two million years ago, the Crescent Terrane slammed into the southwest corner of the island archipelago, adding what would one day be Victoria's bedroom communities of Colwood, Metchosin and Sooke, as well as the Olympic Mountains across the Juan de Fuca Strait in Washington State. Geologists call the volcanic bedrock in this southwestern corner of Vancouver Island the Metchosin Igneous Complex. It's best viewed on the beach west of Albert Head, where a small marine fossil bed crowded with shells can be seen. The shells were once believed to be 150 or even 200 million years old—of Jurassic or Triassic age. Geologists now know that they are far younger, perhaps forty-five or fifty million years old.

The latter collision line, the Leech River Fault, is today one of the island's most prominent valleys. Narrow and steep-sided, it extends from Sombrio Point on the Strait of Juan de Fuca to Esquimalt Lagoon (see page 258).

The volcanoes resulting from these violent impacts have yielded copper (near Duncan) and gold in many places, including the streams near Leechtown. The rock of the ancient terranes have also provided coal and limestone, which have been mined and quarried in many places. One of the best known such quarries is the one that now features lilies, lupins and lavender, rather than limestone—the world-famous Butchart Gardens.

For millions of years following the creation of Vancouver Island, time and weather worked on the landscape, eroding it on the surface and creating caves and caverns in the karst bedrock.

For the past two million years, these long and leisurely processes have been galvanized by the effects of glaciation. As the great sheets of ice waxed and waned, the island was sculpted and smoothed, sand and sediments were excavated and deposited and clay and peat were laid down in many places.

Though it seems that the area around Victoria was less affected than other parts of Canada during the last glaciation—the Wisconsinan—and possibly during other, earlier glacial incursions, the surface bedrock in a number of places makes it clear that the southern tip of the island did not entirely escape the great sheets of ice.

Nature, and of course human activity, continue to change the landscape. In fact, the periods of continental collision aside, it might be said that the Victoria region has been altered more in the past 150 years than in the previous 10,000. Despite all that, it is still possible to see the past here, if only one is prepared to look.

Source: Sattellite Image, Southern Vancouver Island, BC, http://ccrs.nrcan.gc.ca/resource/tour/41/index e.php, Natural Resources Canada, 1999. Reproduced with the permission of the Minister of Public Works and Government Services Canada, Courtesy of Natural Resources Canada, 2006.

Saanich Peninsula

ALMOST EVERY long-suffering inhabitant of Canada's other provinces and territories has had one: a smug New Year's email from someone in southern Vancouver Island boasting about daffodils in bloom or the imminent annual February Flower Count. And as hard as it might be to swallow, it's possible that residents of Victoria and its outlying regions might actually be historically justified in their annual bluster. In fact, scientists now believe that even during the height of the Wisconsinan glaciation, the Earth's most recent deep spell of global cooling that many call "the Ice Age", the southernmost reaches of Vancouver Island were unglaciated. While most of Canada lay entombed beneath up to three kilometres of ice, climatologists now believe that the Saanich Peninsula and what is now Victoria may indeed have had the Pleistocene equivalent of flowers in February.

Though the Cordilleran ice sheet did eventually cover all but the western edge of Vancouver Island, that glaciation did not happen until very late, perhaps 14,500 years ago, and—at least in geological terms—lasted for the blink of an eye.

The irrefutable evidence that life thrived here when most of the country was buried by ice comes from the Saanich Peninsula, which stretches more than thirty kilometres north of Victoria. Teeth, tusks, horns and bones of a wide variety of ice age mammals, or megafauna, have all been found here, and radiocarbon dating has proven that southern Vancouver Island provided a glacial refuge for much of the Wisconsinan glaciation.

Other fossils, of both imperial and Columbian mammoths, as well as ancient bison and brown bears, have been found on the Gulf Islands, as well as on nearby American islands across the border in Washington State. But the majority of British Columbia's ice age finds have come from gravel pits on the Saanich Peninsula. In 1895, molars of imperial mammoths were found on James Island (see page 234), just east of today's Saanichton, and again three years later, at Cordova Bay near Victoria, where the sea cliffs and gravel pits are still prime places to search for mammoth remains. Also found on the peninsula were the tusks and a foreleg bone of an imperial mammoth that were radiocarbon dated to 17,000 BP; both prove that mammoths thrived here at the height of the last glaciation.

It's significant that these fossils

This mammoth tusk and replica can both be seen at the Royal BC Museum, located at 655 Belleville Street in Victoria, almost directly south of the stately Empress Hotel.

belonged to imperial rather than woolly mammoths, for North America's largest mammoth needed a relatively temperate climate for survival. Standing about four metres or thirteen feet at the shoulder, with sweeping ivory tusks at least as long, imperial mammoths lacked the dense, dark hair and underfur of their woolly cousins, which survived for more than two million years in the world's arctic regions and,

Both photos: Dennis Fast

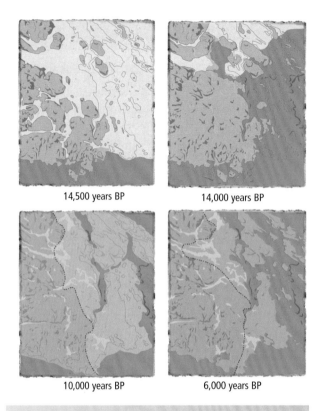

14,500 years BP   14,000 years BP

10,000 years BP   6,000 years BP

On southern Vancouver Island, glacial ice stayed briefly—geologists believe the Greater Victoria area was not glaciated until 17,000 BP—and retreated quickly as the world began to warm. By 10,000 BP, the Saanich Peninsula was covered by grassland with Douglas-fir, while the region to the west was mainly coastal Douglas-fir and Garry oak woodlands. By 6,000 BP, the Garry oak woodlands had moved east; western hemlock forest covered the island's southwest coast.

during glacial intervals, at the edge of the great glaciers. Prodigious grazers, imperial mammoths also needed grass for fodder—and lots of it. Projections based on modern African elephants show that each mammoth would likely have needed to consume at least 180 kilograms or 400 pounds of grass and leaves daily.

Canadian scientist E.C. Pielou contends that 18,000 years ago, which many scientists use as a benchmark for the height of the Wisconsinan glaciation in much of Canada, the ice front was approximately the latitude of Strathcona Provincial Park. If that was the case, it's not surprising that imperial mammoths might be filling their king-sized stomachs more than 200 kilometres south on the Saanich Peninsula. Warmed

by the Strait of Georgia to the east and Juan de Fuca Strait to the south, the peninsula and Victoria just south of it were glacial refugia for a remarkable length of time.

More recent recoveries include another that confirms the image of a parklike grassland about 17,000 BP. The partial skull of a helmeted muskox was found in a gravel pit on the Saanich Peninsula. Now extinct, these tall, lean muskoxen, with flaring horns fused to form a massive helmet, were adapted to warmer conditions than their tundra cousins.

With much of the globe's water locked up in the enormous continental ice sheets, the sea level was 100 metres or more lower than it is today and in many places the Saanich Peninsula would have stretched far out to sea. These huge flood plains in the relatively shallow Strait of Georgia allowed the mammoths, mastodons, muskoxen, bison and other ice age or quaternary species to reach Vancouver Island. In a world that was largely enveloped in ice, the smell of springtime grasses borne on the prevailing westerly winds might very well have lured the megafauna across the remaining narrow sea channels. Fossil evidence shows that imperial mammoths were inhabiting the lower Fraser River valley near Chilliwack about 22,000 years ago and one can easily imagine their descendants wandering west, following their noses, as their homeland cooled.

When the region was finally glaciated—and striations or grooves in the rock just west of the B.C. Legislature, and below the sea cliffs

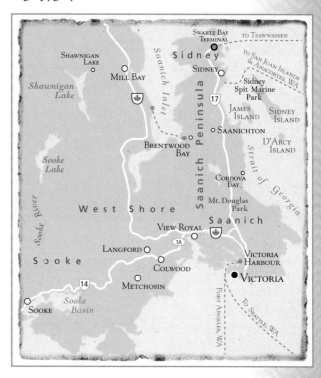

Helmeted muskoxen were well equipped to fend off most predators, but Vancouver Island's changing climate proved deadly.

Amanda Dow

off Dallas Road prove that the ice did eventually reach Juan de Fuca Strait—glaciers elsewhere had already begun their rapid retreat. Pushed from their final sanctuary, mammoths and muskoxen disappeared forever from Vancouver Island. Though they survived the waxing and waning of the ice elsewhere in North America, including nearby Washington State, it seems that conditions—perhaps rising sea levels or winter temperatures that stayed uncharacteristically cool for hundreds of years—prevented their return to the island.

By 14,000 BP, the ice had retreated from the region and it took little time for vegetation to recolonize the barren ground. By 12,000 BP animals adapted to a warmer climate, including *Bison bison antiquus* roamed the peninsula. These ice age bison, with a broad, arched cranium and horns that stretched horizontally from the skull, were descendants of bison populations that had migrated south as the Northern Hemisphere cooled. Later, as the great ice sheets melted, they moved back north. Genetic data show that modern bison are also descended from populations that spent the last glaciation south of the ice.

The first of two bison discovered was unearthed in the autumn of 1987, during the enlargement of an irrigation pond in North Saanich. Excavated more than a metre below the surface, the almost perfectly preserved skull lay amidst freshwater shells in a layer of peat. Dubbed Bison George, it was dated at 11,750 BP, and marked the first time post-glacial bison were found in Canada west of the Cascades. It was followed within a year by the discovery of a second, even more perfectly preserved skull. Found by Grant Keddie, curator of the Royal B.C. Museum, it lay nearly four metres below the surface of a bog near Beaver Lake. Both skulls were smaller than *Bison antiquus* specimens found on the Great Plains, leading scientists to believe that island conditions were less than optimal.

Other large mammals have reacted similarly to island environments. Enticed over a narrow seaway to California's Santa Catalina Islands by the scent of flowering grasses and trees, woolly mammoths ultimately evolved into tiny 1.3-metre or four-foot high pygmy mammoths.

James & Sidney Islands

LYING IN HARO STRAIT, just east of Saanichton on the Saanich Peninsula, James and Sidney Islands are examples of geology in motion. James Island features steep sand cliffs along its southern end; in fact, what appears to be its south shore was once the middle of the diamond-shaped island. The outline of the rest of the island can be seen in the extensive area of shallows on the floor of the sea.

Where did the rest of James Island go? What happened to its southern half? According to the Geological Survey of Canada, strong southeasterly

Peter St. John

winds eroded the glacial sand and gravel and redeposited them along the east and west shores of the island.

In a similar way, Sidney Island is on the move; its southern sediments have been moved from their original position to create Sidney Spit, which is today a provincial marine park.

Viewed from a spit on Songhees land on Saanichton Bay, the newly deposited sands of James Island can be clearly seen across the narrows.

Saanich Inlet

THE LARGEST FJORD on Vancouver Island's east coast, Saanich Inlet is ruggedly beautiful and geologically fascinating, if rather insipidly named. Fjords (the word is Old Norse and is akin to Scottish "firths") are steep-sided, U-shaped valleys that have been eroded by glaciers and then filled by the sea.

Along its twenty-one kilometre length, Saanich Inlet is remarkably varied. From the gentle beaches of North Saanich, it slices south, almost bisecting Vancouver Island. Jogging slightly west two-thirds of the way down, it becomes Finlayson Arm, which is so steep and craggy that builders of the Island Highway were forced to climb more than 335 metres to pass along its western side. Like many European fjords, it is deeper inland than at its mouth, and below the waves it has at least two fascinating stories to tell.

The first is a tale of a devastating flood. A coring project undertaken by the Geological Survey of Canada in 2001 has revealed that about 10,000 radiocarbon years (or 11,800 calendar years) ago, a catastrophic flood roared down the Fraser Valley, across the Strait of Georgia and into Saanich Inlet. It was triggered by the sudden release of an enormous amount of impounded meltwater, likely from a glacial lake in the lower Thompson Valley (see page 105), swamping at least parts of Galiano, Mayne, North Pender and Saltspring Islands.

A world that has witnessed the devastation wreaked by tsunami flood waters in Asia will have no trouble imagining the trees, wildlife and perhaps humans that were swept away as a wall of water, silt and mud roared down the Fraser Canyon and lower valley and across the strait.

Though there are no first-hand accounts of this devastating flood, the evidence is clearly written in the deep clay on the floor of both the strait and Saanich Inlet. Here, in sediment grains that bear the characteristic signature of a flood, scientists were able to identify pollen known to have originated in the Fraser Valley, including lodgepole pine, which was abundant in the lower valley at the time.

The second story of Saanich Inlet is the tale of a dead zone, not the kind one meets in horror movies, but rather an anoxic zone. This deep-water, oxygen-starved region is a result of the shallow sill that lies at the mouth of the inlet, about seventy metres beneath the surface. By restricting the exchange of water between the fjord and Georgia Strait, the sill creates an area below 100 metres deep that is almost completely deprived of oxygen. This makes the fjord bottom uninhabitable, but also preserves layer upon layer of organic matter that would normally be consumed by bacteria. The result is an almost perfect calendar of the distant past. Using modern coring techniques, scientists can compare annual layers, or varves, to the tree rings of 12,000-year-old trees preserved beneath a nearby lake to create a clearer picture of the past.

Peter St. John

Seen from the top of Mount Newton, Saanich Inlet separates the Saanich Peninsula from the heights of the Malahat in the distance.

235

Finlayson Point

JUTTING, THUMB-LIKE, into the Strait of Juan de Fuca, with steep-sided cliffs rising from the sea on all sides but one, Finlayson Point at the south end of Beacon Hill Park is a natural site for a fortification. For hundreds of years, it served just that purpose, as one of a chain of defensive sites that were strung, like pearls on a necklace, around the south coast of Vancouver Island.

Traces of the stockaded village at Finlayson Point can still be seen. From the beach below, one can trace in the cliff wall the profile of a deep protective trench that once cut across the neck of the peninsula. And along the sea wall, on the south side of the site, are mounds and depressions where buildings once stood. But until recently, details on who built the fort, how old it might be, or how long it was occupied have eluded even the experts. Now, thanks to extensive detective work by Grant Keddie, curator of archaeology at the Royal BC Museum, it seems clear that this and other defensive sites were built by the ancestors of the local Coast Salish people before the Norman conquest of England.

Radiocarbon dating has shown that the fortification on Finlayson Point, which is in traditional Songhees territory, was occupied

Both photos: Peter St. John

Jutting into Juan de Fuca Strait, Finlayson Point was a perfect site for a fortified village. With steep cliffs rising from the sea, it served as a stronghold and safe haven for centuries.

almost 1,100 years ago. For almost two millennia before that, the Coast Salish peoples had increasingly built their villages along the island's coastline. People had always lived along the coast, of course, but this move to more exposed coastal areas that were poorer in resources may have been the result of increasingly intense specialization in marine resources. This led to the creation of fortified communities at strategic locations along the coast.

As for how long these coastal strongholds were used, a series of excavations at Finlayson Point convinced Keddie that it was regularly occupied for centuries, until, according to oral history, its inhabitants fell victim to the ravages of foreign disease.

What is less certain is whether the site was occupied permanently, or seasonally, or used only when needed during times of strife. The first two are certainly possibilities; Finlayson Point is located immediately below the very productive Beacon Hill camas fields (see map and Beacon Hill Park on page 239). Excavating a deep midden, or refuse pile, that had accumulated over the centuries inside the

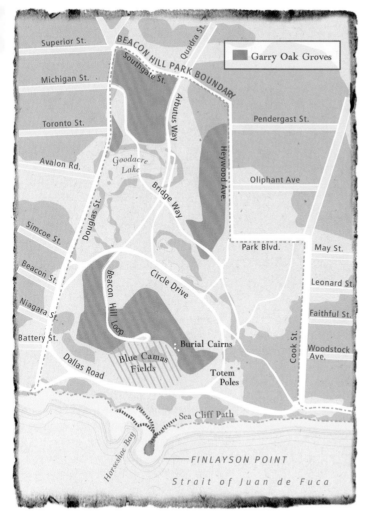

park, the defensive trench was more than two metres deep and at least five metres wide. Serving the same purpose as moats did on medieval castles, the trench was further fortified on the seaward side with a tall, stout palisade; the buildings inside were accessed by a drawbridge.

Nineteen stockaded villages of this kind are known to have existed in the Victoria area, stretching from Metchosin on the west to Cowichan Head on the north. Those that have been excavated all appear to have been created between 1,200 and 1,000 years ago. And many have associated clusters of burial cairns.

In some cases, as at Fleming Bay, north of Macaulay Point, defensive sites or fortified villages were built on top of earlier communities. Keddie found that the shell midden at Fleming Beach went back more than 4,000 years, and below that, deep in the sand, he unearthed a pebble chopping tool, clearly left behind by even earlier inhabitants.

palisade fence, Keddie found the remains of many types of fish, as well as mammal bones and shells. Both the camas fields and the midden point to at least a seasonal occupation of the fort.

Just above the point, in full view on the southeast slope of Beacon Hill, are a cluster of rock piles, the remains of burial cairns disturbed by early park development. Amateur excavations of the cairns in the early 1900s, had revealed individual burials of both adults and children, but there was no record of the cause of death. Was the stockaded village permanently occupied, perhaps?

Yet the site was a true stronghold. Curving across the narrow peninsula that connects it with the

Weighing all this evidence—fortified villages with large nearby burial grounds, almost invariably next to productive land and sea resources—led Keddie to believe they point to the development of new technology, a changing social organization or both.

"The very nature of a defensive village would demand greater social cooperation for group survival," he wrote in a four-part series entitled 'Aboriginal Defensive Sites'. "The need to restrict one's movements may have made necessary a more specialized focus on some food resources. Possibly reef-netting technology for large scale salmon fishing was a result of those needs."

Further, Keddie believes that this new social organization was hierarchical in nature, an idea that

Clusters of boulders on the hill overlooking Finlayson Point mark the remains of grave sites. They were likely family burial grounds and perhaps a sign of status.

Peter St. John

gains credence from the clusters of burial cairns associated with many fortified sites. "I think it is safe to assume," he writes, "that defensive sites were owned by the wealthier individuals who could mobilize supporters to build and defend them. Defensive sites were a visible sign of status for their owners ... [and] family burial grounds ... would also be a sign of status, indicating the quality of their ancestry."

In short, though half a world away, the fortified villages of the Lekwungen or Songhees people were, in a number of ways, like the medieval French and English castles that existed about the same time.

Getting There: Finlayson Point is located on the shoreline in Beacon Hill Park, just south of Dallas Road. The cluster of burial cairns is on the southeast slope of Beacon Hill, just above the point. Lime Bay is on the north side of the Victoria Harbour, at the south end of Catherine Street, while Fleming Bay is in Esquimalt, just north of Macaulay Point.

REEF NET FISHING *Getting a Jump on the Salmon Season*

For the Lekwungen or Songhees people of southern Vancouver Island, as well as their Straits Salish relatives to the west and east—the T'Souke, Saanich and Semiahmoo of what is now British Columbia, and the Lummi and Samish of the San Juan Islands and northern Olympic Peninsula—reef net fishing was the most important technique for catching salmon, and salmon were, in turn, the most important fish caught.

Anchored to a reef on one end, and held in place with floats on the other, reef nets angled toward the surface in the direction of the salmon migration. Part way along the net, ropes were attached to it. When a school of salmon swam up the net, the ropes were quickly drawn up by fishers in canoes, trapping the salmon.

Woven of cedar bark and twine made from nettles, the nets were often camouflaged with eel grass and other plants. Taking the salmon before they entered the Fraser River or other smaller spawning streams gave the Straits Salish nations a jump of almost a month on other salmon fishers.

Communal net sites were located in several areas, but one of the most productive was the reef that extended from the southern tip of Point Roberts (see page 185), southwest of the mouth of the Fraser River. Here, salmon flowing northward along the coastal flats poured over the reef as they rounded the point en route to the Fraser. Netting positions were so valuable that they were passed down in families for generations.

Beacon Hill Park

Gordon Friesen,
Courtesy of The Friends of Beacon Hill Park

For many, Beacon Hill Park means fields of flowers, clumps of venerable oaks, great blue herons clustered on a conifer and limitless sea views, all—apart from the sea views, of course—deemed to be the legacy of prescient city planners and industrious horticulturalists. While the planners and horticulturalists undoubtedly deserve credit for preserving the park's seventy-five hectares and sustaining its inviting environment, each of these treasures (again, excepting the views) could be found in far greater measure 1,000 years ago. And increasingly vocal groups of citizens believe that the best of the past ought to be resurrected.

Once, the lower slopes of Beacon Hill were so bathed in blue camas blossoms in May that it must have seemed the sea had risen from the shore to immerse the land. This flood of flowers was followed, as the seasons progressed, by golden paintbrush, choco-late lilies, lupins, buttercups and violets. Flitting from blossom to blossom were clouds of butterflies—some forty species of them. And framing this sea of color, ancient Garry oaks grew in profusion, sheltering plants, mammals, birds and am-phibians, many of which are now rare or extinct.

Writing in 1843 about this parklike landscape, which was so different from the dark (and to European eyes, forbidding) forests of much of British Columbia,

Hudson's Bay Company Chief Factor James Douglas called it "a perfect 'Eden' in the midst of the dreary wilderness of the North". He further mused that it might have "dropped from the clouds into its present position".

But this was not simply Nature's work; the magnificent May fields of blue camas, and their much sought-after bulbs, for example, were the result of long-held agricultural practices

Peter St. John

These slopes of Beacon Hill were once cloaked in May with blue camas blossoms (inset). In the fall, as shown above, the Lekwungen women would return to the field to harvest the bulbs; some were the size of tangerines.

—careful species selection, controlled burning, and the painstaking clearing of rocks and brush—on the part of the Lekwungen or Songhees people.

Because the bulb of the deadly white camas is virtually indistinguishable from its edible cousin, eliminating the aptly named death camas bulbs had to be done during flowering. Having culled the undesirable plants, the Lekwungen would return to their plots in the fall and, lifting small sections of soil,

© Kerry Lange

239

Allan Taylor

Garry oaks, now endan-
gered, produce sweet
acorns and durable wood.

The island's ancient history
can be seen in the volcanic
rock at the summit of
Beacon Hill.

remove the larger bulbs and replace the earth about those that remained. The fields were then burned. The fires removed the dry grass and shrubs and acted as a fertilizer, promoting new growth during the winter rainy season. The young grasses, in turn, attracted deer and elk, whose droppings further fertilized the land.

This cycle produced huge harvests of blue camas bulbs. Steamed in large pits, the bulbs became soft, dark and sweet—comparable in taste, apparently, to a baked pear—making dried salmon considerably more palatable. The surplus was traded to the Nuu-chah-nulth or other neighbors.

These practices—regular burning, aerating the soil and annual culling—appear to have dramatically increased the size of the remaining bulbs. Recent studies at the University of Victoria have demonstrated that in a similar environment, blue camas bulbs can grow as large as tangerines. In other

words, the Lekwungen had developed sophisticated horticultural practices centuries before Europeans arrived in British Columbia.

The Garry oak (or, as Americans know it, Oregon white oak) woodlands were similarly assisted by native agricultural practices. Regular controlled burns kept competing shrubs and trees at bay, but little affected the oaks, with their thick bark and deep roots. Slow-growing, with small, sweet acorns, British Columbia's only native oak can reach a height of twenty metres or more. For millennia its durable wood (along with that of ocean spray, a shrub dubbed "iron-wood") was used to make sticks for harvesting plants, roots and clams, as well as for fuel. Acorns were roasted or steamed.

Part of a hardwood forest that gradually covered much of southeastern Vancouver Island and the Gulf Islands during the Hypsithermal, the period of global warming between 9,000 and 6,000 BP, Garry oak woodlands reached their greatest extent about 7,000 years ago, before diminishing as the global climate cooled.

Coastal peoples valued these open woodlands, with their flower-filled

E. Sandys, National Archives of Canada, C-013843

Perched one-legged on the upper limbs of a huge cedar in Beacon Hill Park, great blue herons seem as much at home as they do in the wetlands far below. The city's "heron cam" keeps an electronic eye on the birds' activities in the park.

Darren Stone, *Victoria Times-Colonist*

meadows, deer and elk grazing in the dappled shade, and birds nesting among the branches, and preserved them for millennia. As indicated earlier, Europeans also found this ecosystem inviting, but rather than protecting it, in less than two centuries, it has been all but destroyed by settlement and urban encroachment.

From an estimated area of nearly 1500 hectares in the Victoria area in 1800, Garry oak woodlands covered just twenty-one hectares 200 years later—a loss of 98.5 per cent. The north slope of Beacon Hill and the north end of the park together preserve Canada's only extensive relic of Garry oak woodlands, with several trees estimated to be more than 400 years old. Elsewhere, small groves can be found on private property on southeastern Vancouver Island and the Gulf Islands, with isolated trees in the Fraser Valley. In the U.S., patches of oak woodlands are found in the interior coastal regions of Washington, Oregon and northern California.

Great blue herons prefer conifers to oaks, and have made southern Vancouver Island a nesting site of choice for millennia. Though reduced in numbers, they can still be seen in Beacon Hill Park. Gathered in noisy heronries, or poised, motionless, in a coastal marsh or inland pond, these magnificent birds are symbols of our natural environment and finely tuned indicators of environmental problems. Today, as a result of constant encroachment on their nesting territory and pollution of the coastal waters, these elegant, long-lived birds are identified as a species at risk.

Blue camas is much harder to find in Beacon Hill Park. What was once among the most spectacular camas fields in North America, cloaking the area between the Beacon Hill's south slope and Juan de Fuca Strait, is today an off-leash dog park. Despite the tireless efforts of the Friends of Beacon Hill Park, native plants are being replaced by hardy exotics able to withstand regular mowing and trampling.

Things may be changing, however. Increasingly aware of the importance of Garry oak woodlands— one of Canada's four most endangered ecosystems— and the historic role of the camas fields, Victoria's Parks Department has begun working with a community group to plant a restoration plot beside the park's southeast woods, while the Friends of the Park are creating a demonstration area near the corner of Douglas and Southgate Streets.

So, while it is difficult today to envisage this unique landscape as it was 1,000 years ago, perhaps a growing awareness and the combined efforts of many will return at least parts of this once-bountiful place to a semblance of its past beauty.

Getting There: Beacon Hill Park is south and slightly east of Victoria's Inner Harbour. Follow Douglas Street south of the Empress Hotel to Dallas Road (Kilometre 0 on the Trans-Canada Highway). The park is on your left; several roads lead into it.

Peter St. John

E. Sandys, National Archives of Canada, C–024165

The Gorge Waterway

SLICING DIAGONALLY northwest between the municipalities of Saanich and Esquimalt, the Gorge Waterway is a picturesque tidal inlet featuring a lovely little reversing falls. Though crossed by several busy bridges, on the whole it's a quiet place, lined by parks and frequented by walkers, picnickers and kayakers. But this charming backwater has a long, fascinating history.

About 12,000 years ago, when bison were wandering the Saanich Peninsula to the north, the Gorge Waterway was wider and deeper, reflecting the higher sea levels that resulted from the rapidly melting glaciers to the north and east, and edged by forests of pine and alder. As the massive ice sheets retreated and global temperatures climbed, sea levels dropped, until about 9,000 years ago they were lower than they are today. For perhaps four millennia, during a period of global temperatures that were much warmer than today's, the upper waterway became a series of shallow lakes and ponds linked by streams and bogs. Then, about 5,000 years ago, cooler weather returned, restoring sea levels to approximately where they are today.

Earlier people undoubtedly hunted in the forests and marshes, but by 5,000 BP people had established regular camps along the waterway. Evidence of one of these was discovered just above the falls, when a large shell midden was exposed during construction under the

Tillicum Road Bridge. Below a thick layer of oyster and clam shells and herring bones lay a dark layer of soil and charcoal. When archaeologist Grant Keddie had a sample from this layer radiocarbon dated, it came back with an age of 4,120 years. Below it was more than a half-metre of fire-broken rocks, along with sparse shell and bone. Clearly, this was a place people had lived for more than five millennia.

Similar camps were dotted all along the waterway, for in addition to the shellfish, both fresh and salt water fish thrived here—including herring, several species of salmon, cutthroat trout, rockfish and flat fish. In the 1890s, archaeologist Harlan Smith noted that on the north side of the Gorge from Tillicum Bridge to Admiral Bridge was a "continuous shell ridge, composed largely of pure shell material". Unfortunately, most of this ridge is covered today by the sea wall, walkways and landscaping of the Gorge Waterway Project. However, a large shell midden, studded with artifacts, was uncovered at Kosapsom, a village that served the Coast Salish people for thousands of years on the north side of the waterway just east of Admirals Bridge.

Excavating extensively in what is now known as Craigflower Park, archaeologists found obsidian (or volcanic glass) microblades and nephrite (or jade) adzes for woodworking, as well as beautifully-faceted, medium-sized ground slate points. The slate points, which would have been used to tip harpoons, were immediately identifiable. They belonged to the Locarno Beach culture, which thrived in many places in the Gulf of Georgia beginning about 3,500 years ago. Though coastal peoples had been at home on the sea for thousands of years, over time, the weapons they used to hunt large fish and sea mammals, including sea lions or even porpoises, became ever more specialized.

In most places, the Locarno Beach culture lasted about 1,000 years, giving way to what archaeologists call the Marpole culture about 2,400 years ago. At Kosapsom, however, the people apparently continued to use the older Locarno Beach tools and implements for 400 years after they had gone out of

Relatively unchanged in the century since the watercolor at left was painted, the Gorge still reveals some of its long history.

Photos: Peter St. John

built their houses, made their canoes, wove their mats and created baskets, clothing and carvings of great beauty and symbolism.

In short, the quiet, apparently out-of-the-way Gorge Waterway is in fact a fascinating time capsule of life in the region for the past 5,000 years. Today, though local archaeologists would like to see much more be done with it, the ancient site below Tillicum Bridge has at least been caged in to allow visitors to view the deep shell midden, which rises above the Gorge, just below the falls. Here, the history of the place resonates, as it does upstream.

Though the excavations at Craigflower Park are no longer visible, the past is clearly present, assisted by nearby Craigflower Schoolhouse, which served one of Vancouver Island's first farming communities. Across the waterway is Craigflower Manor, a Georgian manor house constructed in the Hudson's Bay Company's post-and-beam style for Kenneth McKenzie, who supervised the farm.

Construction crews working under the Tillicum Bridge (above) found a deep shell midden (at right), which proved to be a time capsule of life along the Gorge. Though caged behind chain link fencing today, part of the long history of the waterway can be clearly seen.

style elsewhere. Was this a case of scientists stumbling on the archaeological equivalent of an Amish settlement, with its reliance on buggies and wagons in car-crazy twenty-first century Canada? Or was there a problem with the C-14 dating. As Keddie writes, "more excavations are necessary … to solve this problem".

Above the layer of out-of-synch artifacts was a thick layer of bone spear points, harpoon heads and awls, along with antler wedges that were quickly recognized as belonging to yet another phase of development, the Gulf of Georgia culture. These were tools that many cultures on both sides of the Georgia Strait began using about 1,650 years ago and, with improvements and variations, continued to use until Europeans arrived in British Columbia.

Like their descendants, these ancestors of today's Lekwungen (or Songhees) and Esquimalt peoples were superb woodworkers, and particularly adept at working with cedar, which grew in abundance over most of the island. From this remarkable wood, they

Getting There: The Tillicum Bridge shell midden is located beneath the bridge on the south side of the waterway. On the north side, a walkway runs along the Gorge west to Kosapsom, which is located at Admirals Bridge, southeast of the junction of Admirals and Gorge Road.

Western Redcedar

FOR AT LEAST the last 4,000 years, the lives of the peoples of the Pacific northwest have been inextricably intertwined with the western redcedar. This remarkable tree, which can soar to heights of seventy-five metres on the coast, reach a diameter of nearly six metres and live for more than 1,500 years, has been used for almost every purpose imaginable, from magnificent whaling canoes, monumental longhouses and soaring totem poles to ceremonial masks, clothing, rope and even medicines. Little wonder that the Kwakwaka' wakw people call it "the tree of life" or that B.C. chose the redcedar as its provincial tree in 1988.

In 1860, when this lovely watercolor was painted, longhouses spanned the length of Songhees Point, on the west side of what is now Victoria Harbour.

Yet, despite its remarkable longevity, wonderful aroma and arrow-straight grain, western redcedar is not, in fact, a cedar. True cedars are found in the Mediterranean and resemble some of Canada's firs. Instead, B.C.'s provincial tree is an arborvitae, one of only three on Earth; the others are eastern white cedar and Chinese arborvitae. (Arborvitae, by the way, also means "tree of life".)

Soon after it colonized suitable regions of B.C.—the low or mid-elevations along the coast from southern Vancouver Island to the Alaska border, as well as the Columbia Mountains, East Kootenays and occasional sites in the Thompson Okanagan—many cultures realized its almost magical properties. Bark could be stripped or even planks taken without killing the tree (see page 288); the bark would light during a rainstorm (an obvious advantage) and the wood lasted for decades.

It may have been the way the wood splits from either a living or a fallen tree that inspired the first longhouse; certainly redcedar's long, straight grain made cutting panels for large multi-family homes not only possible, but relatively effortless, considering the tools at hand. The result was magnificent longhouses that became artistic and cultural statements.

Because of its enormous importance, redcedars were treated with great respect. Children learned early how to strip bark, or chose just the right tree for a canoe, a house post or a totem pole. Even today, prayers are often said before the bark, the wood or the tree itself is taken. The late Haida artist Bill Reid composed the following:

> Oh, the cedar tree!
> If mankind in his infancy
> had prayed for the perfect substance
> for all material and aesthetic needs
> an indulgent god could have provided
> nothing better.

DISTRIBUTION OF THE WESTERN REDCEDAR

This pictograph of what appears to be a sea lion on Aldridge Point in East Sooke Regional Park is accompanied by a legend. Apparently a sea creature was responsible for the deaths of so many paddlers on Becher Bay that the remaining members of the community were afraid to venture out onto the water. Faced with starvation, the village was saved when a spirit man caught the sea monster and turned him to stone on the point.

The West Coast Highway

Skirting the southwestern end of Vancouver Island,

the West Coast Highway introduces travellers

to a buffet of archaeology and geology,

all in a picturesque package.

Yᴇᴛ ᴍᴀɴʏ ᴏꜰ ᴛʜᴇ ꜱɪᴛᴇꜱ can be visited in an afternoon. Almost within shouting distance of downtown Victoria, Esquimalt Lagoon is a birders' paradise that has been home to early British Columbians for nearly 8,000 years. A few kilometres farther west, and south of Highway 14, is Race Rocks. Here, nine tiny islands top an undersea ridge that has been recognized for millennia by sea mammals, sea birds, migrating flocks and humans as a place of great bounty.

Nearby Sooke is a community that, like so much of B.C., is growing by leaps and bounds. Yet it retains a small-town charm, as well as several must-see sites, including the Sooke Potholes, located five kilometres north of the highway. Dotted with tiny beaches, secluded swimming holes and miniature waterfalls, this is a lovely place to spend an hour, or several days, and also a rare example of the impact of a "warm glacier" on a sandstone landscape.

Be sure to stop at the Sooke Museum and Visitor Centre, which boasts a wide collection of artifacts from the region's long and fascinating aboriginal history, including a 200-year-old Nuu-chah-nulth whaling harpoon, as well as more modern objects.

West of Sooke, the highway bends north as it begins its scenic track to Port Renfrew. Along the way, it passes French Beach Provincial Park, an excellent place to spot whales in early spring or lie on the sand and splash in the waves in mid-summer.

Continuing northwest, Jordan River, popular with surfers, windsurfers and kayakers, boasts a campground on the beach overlooking Washington's Olympic Mountains. A couple of kilometres farther is China Beach and Juan de Fuca Provincial Park and Marine Trail, established by the Commonwealth Nature Legacy following the 1994 Commonwealth Games.

Winding forty-seven kilometres through old-growth forests, along beaches and, aided by suspension bridges, across canyons and rivers, it follows a Nuu-chah-nulth trail past Sombrio Beach, where the Leech River Fault marks the impact zone between the Crescent and Pacific Rim Terranes, to Botanical Beach, one of the richest tidal areas on the west coast. The trail can be accessed at both ends as well as in the middle.

Jack Most:

Esquimalt Lagoon

ESQUIMALT LAGOON means flocks of birds—and birdwatchers. Perhaps it always has. An important staging area for migrating waterfowl, which congregate spring and fall near the 'lobe' portion of this ear-shaped lagoon, this is a place that has provided bountiful harvests for people for thousands of years.

Along with the ducks and geese that gather in the fringe marshes along the western shore, the diverse habitats of the lagoon draw many other species of birds. To the west and north are upland forests, riparian woodlands, Garry oak meadows and wetlands, while to the east and north are dune grasses, beds of eelgrass and both salt and fresh water.

Each ecosystem supports its own community of life. The eelgrass, for example, provides cover for cutthroat trout and young coho and chinook salmon. And the channel at the lagoon entrance—there were once two, for what is now known as Cobourg Peninsula was once a sand bar open at both ends to the sea—where tidal currents wash over the gravel bars, provides ideal habitat for butter and littleneck clams, mussels and oysters. Elk were once found here and deer can still be seen in the forests to the north and west, where Fort Rodd Hill NHP and Royal Roads

University now stand. And all this food and shelter draws songbirds, flickers, loons, herons, hummingbirds, grebes, mergansers, cormorants, scaups and gulls, among many other species.

Archaeologists have located at least four large shell middens—or ancient refuse heaps—on the university grounds, but it was not until new city sewers were installed that proof of the lagoon's antiquity was revealed.

Excavating a trench in February 2003, construction workers came across two burial sites, including a cairn burial typical of those between 1,000 and 1,500 years old. More was to come. Deep beneath layers of

Fisgard Lighthouse, on Esquimalt Harbour, opened in 1860. It was the first lighthouse on Canada's Pacific Coast; today, it's a National Historic Site. Inset: Esquimalt Lagoon.

Photos: Peter St. John

Ancient artifacts unearthed

■ Artifacts as old as 8,000 years found during power-line digging. Two complete skeletons as well as others discovered.

middens, ancient beaches and peat, under the corner of what is now Ocean Boulevard and Lagoon Road, archaeologists came upon wooden artifacts, including a woven basket, four fish hooks and two wedges that, prior to radiocarbon dating, archaeologist Peter Dady believed might be 8,000 years old.

Wood of that age is rarely preserved, but as in the case of the Danish "bog people", the site's anaerobic state had almost perfectly preserved these ancient artifacts. Dady was even hopeful that further testing on the woven basket might show that it was not made of redcedar, for thousands of years the material of choice for such articles, but of an alternative wood, which might indicate that the basket predated the appearance of the tree on the island perhaps 5,000 years ago

Just northeast of Esquimalt Lagoon lies Esquimalt Harbour, famous for more than a century and a half as the base for first Britain's and later Canada's Pacific Naval Fleet. In fact, sheltered from the sea and edged

with quiet coves and small bays, Es-whyo-milth—"the place of shoaling waters"—has sustained people for millennia.

More than 1,000 years ago, the harbor was the site of at least three of an estimated nineteen defensive sites in the region (see Finlayson Point on page 236). Some were clearly much more than fortified bases. Excavations of the 2,800-year-old Maplebank Site on Songhees land on the east side of the harbor have produced magnificent examples of artwork.

In 1933, when visiting a British colleague who was working at the site, Dominion archaeologist Harlan I. Smith unearthed a beautifully detailed elk antler carving of a figure wearing a headdress and facial markings similar to tattoos worn by Cowichan women. Despite the tattoos, RBCM curator Grant Keddie believes that the figure may be a shaman, for the positioning of the hands are consistent with shamanistic ritual. As for its age, one can only guess, for the carving, unfortunately, has disappeared.

Keddie recovered another elk antler carving while excavating at the same site in 1976. This was clearly a comb, though for dressing human or dog hair—which was used

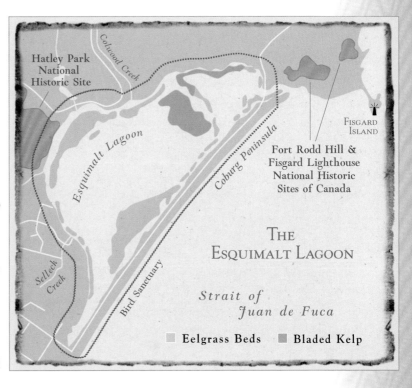

Hatley Park National Historic Site

Colwood Creek

Esquimalt Lagoon

Selleck Creek

Bird Sanctuary

Coburg Peninsula

Fisgard Island

Fort Rodd Hill & Fisgard Lighthouse National Historic Sites of Canada

THE ESQUIMALT LAGOON

Strait of Juan de Fuca

■ Eelgrass Beds   ■ Bladed Kelp

in weaving blankets (see page 139)—or both, is not clear. About the comb, Keddie has written, "The animal-like head, human-like figure and the posture of the figure suggest that it may represent a transformation being, that is, a supernatural creature in transition between animal and human forms. Such transformations were enacted by masked dancers in the historic period."

Carbon dating revealed the comb to be about 1,300 years old.

Getting There: Esquimalt Lagoon can be reached by branching left off Highway 1A (the Island Highway) onto Ocean Boulevard. From downtown Victoria, go north on Douglas Street, which becomes Highway 1. Continue for about one kilometre, then take the right-hand exit toward Colwood. Passing beneath Highway 1 and the railway, continue over the bridge beside the Six Mile Pub and travel up the hill, moving into the left lane. At the lights immediately past the Juan de Fuca Recreation Centre, turn left onto Ocean Boulevard. Continue for .4 km, angle left again toward Fort Rodd Hill, but continue on Ocean Boulevard past the Fort Rodd entrance to the lagoon, which is open to the public. Members of the Esquimalt Lagoon Stewardship Initative ask that visitors respect the lagoon environment, its wildlife and plant life, and take care not to trample plant or sea life. Dogs must be leashed.

John Douglas Sutherland Campbell / National Archives of Canada / C–019327

This view of Esquimalt Harbour was done in charcoal by the Marquis of Lorne, then Canada's governor general, during a cross-country tour in 1881.

Race Rocks

OFF THE SOUTHERN TIP of Metchosin, just west of Victoria, an undersea ridge pokes above the rushing waters of the Strait of Juan de Fuca, creating a place of unmatched biodiversity that has been recognized for centuries. In the Clallum language of the Coast Salish people, this scattering of islets was called XwaYen (pronounced shwai'yen) "the fast-flowing waters".

densely populated ecological regions in Canada. A thriving community of sponges, anemones and soft corals thrives in the sub-tidal and intertidal areas; crowds of barnacles, urchins and sea stars adorn the underwater ledges, and rare rockfish, sculpin and lingcod hide in the rocky crevices and populate the undulating kelp forests.

Above the waves, the nine tiny islands serve as nesting sites for colonies of seabirds and stopovers for thousands of migratory birds.

This is also the largest haulout and feeding area for harbor seals in Canada,

Map redrawn, from Department of Fisheries and Oceans

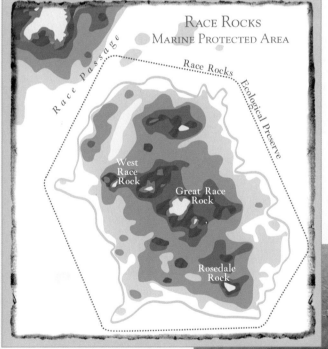

RACE ROCKS
MARINE PROTECTED AREA

Race Passage

Race Rocks

Ecological Preserve

West Race Rock

Great Race Rock

Rosedale Rock

> One of only three tiny undersea areas that are completely off limits to fishing in British Columbia, Race Rocks is among the most densely populated ecological environments in Canada.

and Stellar and California sea lions flock here by the hundreds, fattening up each year for the journey back to California and Mexico. Orcas, today a threatened species, cruise the nearby waters; gray whales,

© Lester B. Pearson College of the Pacific / racerocks.com

The nutrient-rich water, borne on a tidal flow of up to seven knots, and the undersea cliffs, which rise from a depth of twenty fathoms, combine to create one of the most

Dating from 1860, the Race Rocks Light Station is the second-oldest lighthouse in Western Canada. Older by months is Fisgard Lighthouse at Fort Rodd Hill.

Dall's and harbor porpoises are sometimes seen and the endangered northern abalone populates the undersea shelves.

In short, in an area of less than 220 hectares is one of the most ecologically diverse regions on North America's west coast. Little wonder that at least four Coast Salish nations found this a bountiful place and shared its resources year-round.

Recognizing its importance for many species,

resource extraction.

Surprisingly, perhaps, given its proximity to Victoria's sewage outfall and the tanker traffic in the nearby Strait of Georgia, water quality at Race Rocks continues to be high, thanks to the flushing action of the ocean currents. Nevertheless, this remarkable site is vulnerable to oil

Dennis Fast

Race Rocks has been named a national marine protected area, with the potential to be declared a World Heritage Site. Incredibly, however, given British Columbia's enormous pride in its bounteous ocean frontage and the importance its surrounding waters have had throughout its long history, Race Rocks is one of only three tiny areas—totalling less than five square kilometres—that are completely off limits to fishing. The other two are Porteau Cove Marine Park in Howe Sound (see page 199) and Whytecliff Marine Park just west of Horseshoe Bay. Contrast this to the more than 100,000 square kilometres of terra firma in the province that are now protected from most

Harbor seals clearly don't mind a crowd. Today, healthy populations are easy to find along B.C.'s coast, for numbers have rebounded to over 130,000.

spills, nearby commercial and recreational fishing and even eco-tourism.

Getting There: Race Rocks is located just south of Christopher Point, which in turn is south of Metchosin. Several companies offer boat tours of the site. One of its nine islets is the site of Race Rocks Light Station.

The Sooke Potholes

EVEN ON A HOT SUMMER DAY, when the clear, green water of the Sooke River beckons, many have paused to wonder, "How, exactly, were these potholes formed?" The answer lies in a perfect combination of an easily molded material, exactly the right blend of sculpting agents, and a giant-sized sculptor to put the two together. All these things were present 15,000 years ago along the Sooke River. The molding material was the Sooke Formation, easily eroded sandstone interbedded with siltstone and conglomerate—a kind of natural cement. The sculpting agents were a combination of sand, pebbles and boulders, which might have come from the conglomerate, or may have been brought from the uplands to the north. And the sculptor was the glacier that flowed down the valley about 15,000 years ago.

Moving south, the ice was very likely what geologist Ian Walker has called a "warm" glacier. Unlike the continental ice sheets (the "cold" glaciers) that enveloped most of central and eastern Canada and moved by slowly spreading, pancake-like, from the tremendous weight at their centres, warm glaciers glided quickly and easily over the rock on a bed of water. Produced by a combination of the glacier's weight and its speed, the sediment-laden water not only enabled the ice to move easily, but also, from time to time, suddenly released and then trapped sediments beneath the ice. Held there, agitated by the glacial water, the debris rapidly eroded the sandstone, polishing the edges of the canyon, while boulders and large rocks swirled in place to form the potholes.

It's something to think about as one leaps from the smooth rock into the deep, cool water, walks the trail along the edge of the river from the lower parking lot, or bikes the Galloping Goose Trail to the campground and beyond: a glacier that came late, moved quickly and didn't linger can be thanked for creating one of Vancouver Island's most popular swimming spots.

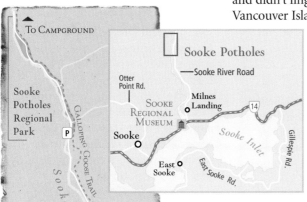

Getting There: Follow Hwy 14 north, toward Sooke. Just past Edward Milne School, turn right onto Sooke River Road, which leads past riverside homes to Sooke Potholes Provincial Park. Two parking lots, with adjacent wheelchair accessible washrooms can be found along the river; trails lead to the various beaches, potholes and viewing points. The campground, operated by The Land Conservancy, is located above the second parking lot. Vehicle access is for the campground only above this point. The Galloping Goose Trail, a popular hiking, biking and equestrian trail through the region, follows the river through the park and beyond.

253

The T'Sou-ke

LIKE THEIR STRAITS SALISH relatives on both sides of Juan de Fuca Strait, the T'Sou-ke have for centuries been a people at home on the sea. Even today, T'Sou-ke paddlers take part in the renewed "tribal journeys", which began with a canoe journey to celebrate the 1989 Washington State centennial. Since then, they have become an increasingly important way of introducing young people from many First Nations on Vancouver Island and the Olympic Peninsula to their ancient cultures and traditions.

T'Sou-ke territory appears to have originally stretched along Vancouver Island's most southerly coast from William Head, east of Pedder Bay, past the bountiful Race Rocks to Becher Bay, and west again to Otter Point and Point No Point, where the Pacific begins to pound in earnest against the shore. This ancestral land was centred on Pedder Bay until smallpox, fur trading and European settlement at the end of the eighteenth and early nineteenth centuries created a swath of death and resulting havoc among many nations across North America. The Sooke Basin, where today's T'Sou-ke Nation is headquartered, was originally Skwanungus territory, according to Cheryl Coull's *A Traveller's Guide to Aboriginal B.C.*

The T'Sou-ke homeland was a bountiful region. Theirs was never a culture wholly based on salmon, like the Coastal and Interior Salish peoples of the Fraser and Thompson Rivers; nor were they whalers, like their Nuu-chah-nulth neighbors to the northwest.

Nevertheless, salmon were important and the T'Sou-ke used sophisticated reef nets to capture schools swimming close to shore as they began their long migration from the Pacific to the rivers and streams inland. Later, after the salmon moved to the mouth of the Sooke River, the T'Sou-ke used weirs across the lower Sooke River to take spawning salmon as they headed upstream.

They also hunted seals and California sea lions. Today, these sea mammals can still be found hauled out on Race Rocks, well within traditional T'Sou-ke territory, but 500 or 1,000 years ago both populated the strait in greater numbers.

The T'Sou-ke also found bounty in the forests, woodland glades and coastal marshes. A wealth of foods waited to be gathered, from hazelnuts (found only on the island's southernmost tip) and Oregon grapes to blackberries (delicious, but dangerous to collect) and bulrushes, which were used to weave mats, capes and bags.

Like many of their neighbors, as well as the Coast Salish nations along the Fraser Delta, the T'Sou-ke depended on the western redcedar for everything from their longhouses to articles of clothing.

Life was more than high adventure and hard work, however. The painting above, which was created for the Sooke Museum, shows the

Whaling harpoons, such as this 200-year-old Nuu-chah-nulth beauty, were often made of western or Pacific yew. Crafted of two long, straight trunks, it was cannily notched to fit together. Known as the "bow plant" for its tough, durable wood, yew had many other uses, from awls and digging sticks to dipnet frames and fire tongs. The harpoon rope, which had to be long and strong enough to hold the whale when it dived, was made of fibre from stinging nettles, woven together for strength. One has to admire the perseverance of the women who collected the plants.

This painting was based on one done from life by Paul Kane, as well as on traditional stories passed on by T'Sou-ke elders.

popularity of *s'la hal*—"the bone game". Players crowded on the shore in teams, each led by a "guesser". As Sooke Museum curator Elida Peers explained in *The Sooke Story*, it was the guesser's responsibility to "correctly guess which of the two concealed bones of each set [was] the unmarked one." Points were won by the opposing side when a guesser missed identifying which of its players held the unmarked bone.

One can imagine the growing excitement, with drums beating, the crowd chanting and items of value on the line.

Getting There: More on life "where the rainforest meets the sea" can be found at the Sooke Region Museum and Visitor Centre, located on northwest cor-

ner of Hwy 14 and Phillips Road, immediately west of the Sooke River Bridge. Open from 9 a.m. to 5 p.m. through most of the year, the centre is closed on Mondays during the winter and for a short period over Christmas. Trails in East Sooke Regional Park invite exploration; one leads to the top of the cliff on which the petroglyph shown on page 246 was carved.

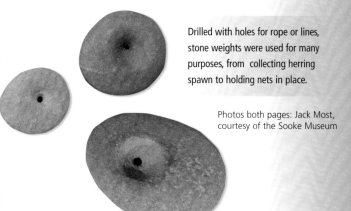

Drilled with holes for rope or lines, stone weights were used for many purposes, from collecting herring spawn to holding nets in place.

Photos both pages: Jack Most, courtesy of the Sooke Museum

French Beach

Amanda Dow

FRENCH BEACH lies at a cultural and ecological crossroads that has been recognized for millennia. Though today deemed to be T'Sou-ke territory, the history of the T'Souke people in the area goes back perhaps two centuries. Prior to that, French Beach belonged to the Pacheenaht, "people of the sea foam". A member of the Nuu-chah-nulth culture, the Pacheenaht First Nation is currently headquartered near Bonilla Point in Pacific Rim National Park. Its traditional territory, however, extended southeast along the Pacific coast past French Beach to Shering-ham Point, where the protected leeward shores of southern Vancouver Island give way to the land that faces the open Pacific—the land of the sea foam.

Like the other inhabitants of Nuu-chah-nulth villages that were strung, like life buoys, along North America's wild, western shore, the Pacheenaht were completely at home on the mighty Pacific; for thousands of years they lived royally off the ocean's endless bounty.

This abundance was particularly evident in the first days of spring, for March brought whales, hundreds of them—humpbacks, gray whales, minke whales and orcas—to the bays of the island's west coast and the estuaries of the Strait of Georgia.

The return of the great whales was cause for celebration and wonder, conservationist Nancy Baron recalled in 1998 in the *Georgia Straight*, "gamboling in the estuaries and lolling at the river mouths". Or, as she quoted from *The British Colonist* newspaper of July 22, 1869, "saucily spouting and sporting" barely off-shore. March and April meant the arrival of herds of barking, bellowing sea lions, bald eagles swooping from the tree branches or gliding above the water, Pacific albatrosses floating above the waves, and clouds of

Getting There: French Beach Provincial Park, which is open year round, is located 20 km west of Sooke, on the West Coast Highway (No. 14). The park has a day use area, a small campground with limited facilities and a one-kilometre walking trail through second-growth Douglas-fir and Sitka spruce. Just 13 km northwest of French Beach is Juan de Fuca Provincial Park and Marine Trail, which were established in 1996 to celebrate the Nuu-chah-nulth and Pacheenaht cultures and protect some of their many ancient village sites, which line Vancouver Island's Pacific shores. Here, too, is Botanical Beach, which is particularly significant ecologically.

French Beach ◄ — 157 km — ► Victoria

Victoria

French Beach

14 Sooke Road

Pacific Ocean

seabirds and shorebirds, refuelling on their annual migration north.

There was a reason for this magnificent spring gathering—herring—specifically the annual spawning of the Pacific herring. Returning each year in numbers that are all but unimaginable today, they arrived in a roaring flood of life, at times bulging two feet out of the water. Heading for the protected inlets, each female would lay as many as 20,000 eggs, which the males fertilized as they were laid. On contact with water, the eggs became sticky and clung to anything they touched—kelp, seaweed, eelgrass … or spruce boughs.

The Nuu-chah-nulth, along with the Makah people to the south and the Haida farther north, also gathered for this outpouring of bounty. Hanging weighted spruce boughs from an anchored frame of slender poles, or weighting fir branches or kelp in a shallow bay, they were easily able to harvest the egg masses that soon covered them. The laden kelp and branches were then hung to dry and the crisp, dried eggs—a favorite food for many cultures—were stored in lidded boxes.

Today the herring are nearly gone. Decades of commercial overfishing has all but silenced the waters that once roared with life. And the great congregations of sea mammals and birds that once relied on this bounty cling to the edge of existence.

One of the few places where the herring, and their accompanying assemblages, may still be seen is at French Beach. Between mid-March and early April, migrating gray whales, otters, seals and sea lions can sometimes still be seen offshore. And slowly, the wisdom of the people who lived for so long in harmony with nature on the Pacific shore, is being heeded by those newcomers who, for more than a century, believed the bounty of the oceans to be endless.

Peter St. John

Opposite: The herring arrived in such numbers they could be literally swept from the sea.

Huge logs and enormous roots litter the beach, proving the power of the Pacific.

The Leech River Fault

HIKERS WHO HAVE TREKKED the Juan de Fuca
Marine Trail along Vancouver Island's southwest coast
have called it "otherworldly". They may be simply
revelling in Nature's splendor, or the feeling of being
completely apart from the normal bustle of Canadian
life, but they are also fundamentally right; the island's
southwestern corner *is* a place apart, the last of the
many "exotic fragments", as the Sooke Museum has
called them, that constitute modern British Columbia.

The truth of this is clear in satellite photos, which
show the Loss Creek-Leech River Fault as a steep-
sided valley running from Sombrio Point diagonally
across the corner of the island to Esquimalt Lagoon.
It's also obvious to geologists on the ground, who have
found that the bedrock of Sooke and areas to the west
consist of fifty million-year-old, blue-black Metchosin
volcanic rocks. Here is evidence of where the Crescent
Terrane—a volcanic island similar to modern Iceland
—slammed onto and under the older rock of Pacific
Rim Terrane about forty-two million years ago. It may
have originated at the edge of the subducting Juan de
Fuca oceanic plate. South of Vancouver Island, a line
of sea-floor volcanoes marches southeast from the
Crescent Terrane to form the San Juan Islands and the
Olympic Mountains across Juan de Fuca Strait.

The Pacific Rim Terrane, by the
way, is another fragment of the island's
exotic collage, having collided with
the much older and larger Wrangellia
about fifty-four million years ago.

The rock in the Pacific Rim
Terrane is older than its neighbor to
the south and consists of slaty schists
—fine-grained metamorphic rocks—
from the Jurassic and Cretaceous
periods. And in these rocks is gold,
which has eroded into virtually every
stream in the Leech River region,
including the Sooke River. Geologists
believe that the subducting Crescent
Terrane created volcanic activity,
which in turn generated "plumbing"
that allowed the gold to be trans-
ported and deposited.

Gold was a big draw for the first
European settlers in the area. The
placer deposits were discovered in the
1860s and over the next two decades
more than 3,000 men were engaged
in placer mining along Leech River,

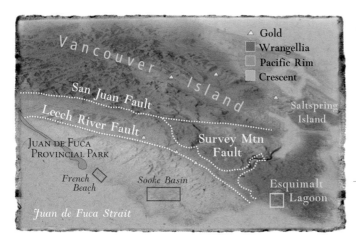

The Pacific Rim and Crescent Terranes
not only created steep-sided valleys *in
situ* when they collided with Vancouver
Island. They were also responsible for
the creation of Washington's San Juan
Islands to the southeast and for the
Olympic Mountains along the coasts
of Washington and Oregon.

Leechtown is gone now; all but the foundations of the old workings have been hauled away, but for a brief time in the region's history, miners crowded the lower Leech River and pitched their tents on its banks.

Photos: Jack Most / courtesy of the Sooke Museum

centred around Leechtown. By 1876, it was estimated that between $100,000 and $200,000 worth of gold had been found.

Most of the gold came from crevices in the bedrock of the river, or from benches along it; nuggets up to 31.1 grams were recovered.

Today, signs of old workings can be seen for about 6.5 kilometres upstream from the Sooke River. And according to celebrated geologist George Dawson, the run of gold turned up the North Fork, then rapidly diminished. Martin's Gulch is also notable for the gold that was found for a distance of two kilometres up from the Leech River.

But Vancouver Island's gold boom is over and its southwestern corner is now known for treasures of other kinds, including trees that rival the largest in the province—it's considered to be Canada's best region for growing trees—and a history of remarkably self-sufficient people—the Nuu-chah-nulth and Pacheenaht (see page 295)—who have lived for millennia along the province's wild Pacific coast, hunting whales on the open ocean.

259

A blazing sunset turns the rock of the Puntledge River to gold. Many feel the fossils found here are even more valuable.

The Island Highway

Though disguised by both its name and number,

the Island Highway (No. 19) is a continuation
of the Trans-Canada Highway.

RUNNING THE LENGTH of the island from "Kilometre 0" at the south edge of Beacon Hill Park (see page 239), it allows motorists good access (if not freeway conditions) all the way to Port Hardy and the Inside Passage ferry terminal at Bear Cove (see page 275). In fact, it is the only paved road that goes the length of the island.

Along the way, it allows access to the Pacific Coast in several places. Those roads almost invariably follow the ancient trails established thousands of years ago by native British Columbians. Two of them are covered in this book. They are the Pacific Rim Highway from Parksville and Qualicum Beach to Pacific Rim National Park (see page 277), and the Gold River Highway (No. 28) from just north of Campbell River through Strathcona Provincial Park to Gold River and Muchalat Inlet off Nootka Sound (found on page 291).

Particularly in its southern reaches, where it has been twinned to expedite the ever increasing traffic, the Island Highway is businesslike and, in many places, not overly picturesque. There are exceptions, of course. One such is Malahat Drive, which climbs to 352 metres as it surmounts the height of land between Victoria and the Cowichan Valley. A viewpoint just north of the summit provides a spectacular view to the east of the Saanich Inlet and Saanich Peninsula and northeast of the Gulf Islands (see page 211).

But adventures abound just off the highway and much of it is written in the rocks. In Nanaimo, 120 kilometres north of Victoria, the region's warm, watery past has been revealed by a series of paleontological discoveries over the past two decades, including a plesiosaur tooth, fossilized buttercups and palm fronds. Nanaimo also is home to Vancouver Island's most magnificent collection of petroglyphs.

A short distance north, the Island Highway goes through Qualicum Beach, home of the Vancouver Island Paleontology Museum (on page 265), which offers rock stories of another sort. Intimate, yet impressive, the museum—which is open from the May long weekend to mid-September—has an enormous collection of fossils that reflect the passions of its founder, regional expert Graham Beard.

And stone is the star of the Horne Lake Caves (page 267). Located a short distance north and west of Qualicum Beach, the park has some of the most accessible of Vancouver Island's more than 1,000 limestone karst caves. Some are open for self-exploration year-round; others feature guided tours that are run

Peter St. John

on a regular basis during the summer season and at specific times during the rest of the year.

Another fifty kilometres farther north is an attraction islanders likely never thought they'd have: a "dinosaur" museum. In fact, the creatures celebrated at the Courtenay Museum and Palaeontology Centre, as well as on fossil tours of the picturesque Puntledge River, are not strictly dinosaurs, but rather plesiosaurs, huge marine reptiles that populated the oceans around Europe and North America during the Mesozoic era.

For those who fancy fossil collecting between stints on the beach, the ferry to Denman and Hornby Islands (see page 222) leaves from Buckley Bay, about halfway between Qualicum Beach and Courtenay, on an

Nanaimo's petroglyphs include figures that are clearly human.

hourly basis (or even oftener, depending on the time of day).

North of Courtenay, Campbell River provides access to Quadra and Cortes Islands to the east and to Gold River and Nootka Sound to the west. And following the Island Highway to its terminus at Port Hardy and Bear Cove offers the potential to explore the magnificent Inside Passage and British Columbia's northwest coast. But that will have to have to wait for Volume II of *In Search of Ancient British Columbia.*

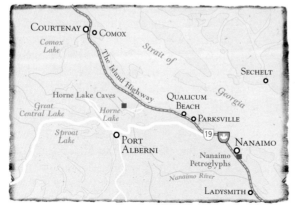

Nanaimo

NANAIMO—the name comes from a Coast Salish word, *snunéy-muxw*, "the meeting place" —can boast of many things, but of particular interest are its rocks, which contain several chapters in the region's long history. Perhaps best known of these are the rock carvings at Petroglyph Provincial Park. Located just beyond the city's south end, near where the Chase River empties into one of the prettiest harbors in B.C., the park includes a remarkable number of magnificent rock carvings including, appropriately enough given the region's early history, what appear to be mythical creatures of the sea.

Painstakingly pecked into the region's local sandstone with chisels and hammerstones made of much harder stone, this concentration of exquisitely rendered sculptures is located on a hill overlooking the harbor and clearly signifies a place of power. Dating petroglyphs is all but impossible, though these clearly speak of a sophisticated culture with the leisure to indulge in the arts. This is hardly surprising, since the Snunéymuxw have been established in this bountiful region for at least the last 5,000 years.

Many of the Gulf Islands, particularly nearby Gabriola Island (see page 220) also boast excellent examples of rock art, but few places in Canada can match the concentration found here. In order to allow visitors to better appreciate this outpouring of art (and even make a copy for themselves) without damaging the originals, the park has an interpretive walkway with informative panels and replica castings.

Other rocks in the Nanaimo area tell of the region's much older history. This "geological biography", as it's been called, includes an enormous palm frond nearly two metres long, arguably the largest fossil leaf ever found in Canada. It was discovered in the mid-1990s during the construction of the approach to a new ferry terminal at Duke Point. Preserved by a sharp-eyed bulldozer operator, this magnificent fossil, which is now at Malaspina College in Nanaimo, was just one of many that were quickly rescued by amateur paleontologists as the construction work continued. Among them were fossilized buttercups, so beautifully preserved that their stamens and pistils as well as the petals could be seen, and a flower that paleontologists Rolf Ludvigsen and Graham Beard have written is "virtually identical to the flowers of *Astronium*, a tropical hardwood of the sumac family". These and magnificently preserved Vibernum leaves and ferns were found in layers of fine Late Cretaceous sandstone and shale in the Protection and Cedar District Formations, which have been dated to between seventy-six and seventy-four million years

Many otherworldly sea creatures are so exquisitely rendered that it's hard to imagine how they could have been made without benefit of modern tools. This, by the way, is a copy.

Peter St. John

263

ago. Together, they have been labelled the "Cranberry Arms fossil suite".

Flowers (and the insects needed to pollinate them) were well established by the end of the dinosaur era. The earliest flowers are believed to have evolved about 145 million years ago, near the end of the Jurassic. To date, the world's oldest fossil flower, which was Early Cretaceous or about 140 million years in age, was found in southern England. Like the fossilized flowers of the Nanaimo region, the English blossoms thrived in the semi-tropical or warm temperate environment that extended over most of the globe.

The fossils near Duke Point (which today is the site of the ferry terminal to Tsawwassen) included both conifers and ferns, indicating a warm temperate climate, similar perhaps to today's mixed forests of southern China and Japan. This may mean that the Insular Superterrane, which included Wrangellia (and therefore most of Vancouver Island and the Gulf Islands), had moved northeast from its point of origin somewhere in the southwestern Pacific, and was beginning to converge with North America.

But the region's rocks tell even older stories. Just northwest of Nanaimo, a quarry near Brannen Lake tells tales about the region's ancient subtropical, seaside past. In 2001, a member of the Victoria Paleontological Society spotted an unusual fossil in Haslam Formation rock, dated to between seventy-eight and eighty-two million years old.

The quarry had previously rendered shark's teeth and ammonites, indicating a watery past, but this was something unusual. Though the fossil broke into several pieces as she pried it loose, it was sturdy enough to be reassembled. After studying it, experts believe it is a plesiosaur tooth, similar to those of the Puntledge elasmosaur (see page 270). The discovery not only marks the most southerly location for this sea creature on Vancouver Island, but also the oldest yet found. The other specimens—including remains on the Trent and Englishman Rivers—were all collected from the Pender Formation, between two and six million years younger than the rock near Brannen Lake.

All this tells a complicated tale about southeastern Vancouver Island during the waning chapters of the Cretaceous. The presence of both lush flowering plants and giant sea creatures indicates that then, as now, the land that underlies Nanaimo sat by the sea. It also shows that the climate was considerably warmer than it is today and that sea levels rose and fell over time, sometimes inundating the land with shallow marine bays and at other times falling to expose seaside deltas and flood plains. Undoubtedly there is more to learn about Nanaimo's distant past and it is likely the rocks that will tell the story.

Getting There: For Petroglyph Provincial Park, follow the road into Nanaimo from the Nanaimo Parkway intersection just south of the city. The park, which is accessed via a pulloff on the east side of the road less than two km from the intersection, features a short, wheelchair accessible trail leading to an interpretive area. The sandstone gallery is a short distance farther along the walkway on a hill overlooking the harbor.

The giant palm frond from the Cranberry Arms is displayed at Malaspina University College, where it is the centrepiece of a newly opened fossil hut. The college is at 900 Fifth Street in Nanaimo.

Vancouver Island Paleontology Museum

at Qualicum Beach

SCENIC, CHARMING even quaint, the community of Qualicum Beach takes its name from the *squal-li*, or "chum salmon", in the Coast Salish dialect of the Pentlatch people, that have spawned for millennia in the Little Qualicum and Qualicum Rivers north of town. The beaches along Vancouver Island's east coast, now part of a Qualicum National Wildlife Area, have also long been known for their bountiful shellfish and, in March, as a place where Pacific herring gather to spawn.

These productive shores draw California sea lions from November to March, as well as northern sea lions and porpoises and harbor seals can be spotted year round. Little wonder the middens of the Pentlatch people have been found in many places along the shore.

But Qualicum Beach also boasts artifacts that reach much farther back into the island's past. The Vancouver Island Paleontology Museum is a testament to the passion and hard work of Graham Beard, who has gathered a huge collection of fossils from the island and beyond, and Tina Beard, who has created illustrations of many of them.

The star of the museum, which is housed in the region's original power buildings, is a remarkably complete skeleton of an adult female walrus. Affectionately nicknamed Rambling Rosie, the fossil was unearthed just ten kilometres north of Qualicum Beach in 1979; found in Early Wisconsinan sediments, it is believed to be between 60,000 and 70,000 years old. The only other walrus specimen to have been found along North America's west coast was dredged from San Francisco Harbor. It has been tentatively dated to about 27,000 BP, but may be much older.

Today, Pacific walruses are found far to the north among the ice packs of the Bering Strait, while their Atlantic counterparts belong to five distinct populations from Hudson Bay north to Hudson Strait and east around Baffin Island and Greenland. That specimens have been found far to the south of this range points to the possibility that a population was isolated south of Beringia during the last emergence of the land bridge between North America and Asia.

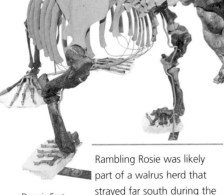

Rambling Rosie was likely part of a walrus herd that strayed far south during the early Wisconsin glaciation.

Dennis Fast

265

Thanks to Graham Beard, the region's resident expert, as well as paleontologist Rolf Ludvigsen and many others, there is much more to see in the museum. One of the most impressive collections is the Vancouver Island ammonites, soft-bodied invertebrates with hard shells. Though ammonites became extinct at the end of the Cretaceous, a relative, the pearly nautilus, survives today. More important, because paleobiologists now know that ammonites first evolved—apparently from cephalopods—about 400 million years ago, and changed over time in a predictable way, and because certain ammonites are associated with very specific periods, they are increasingly being used to date formations using a process is called biochronology.

The museum's enormous collection also include gastropods, with simple spiral shells, and bivalves, like today's clams, as well as shark's teeth. Like their modern counterparts, ancient sharks had replaceable teeth, something that aided in making them superb predators, but also left evidence of their passing. The world's oldest intact shark fossil, which was found in New Brunswick by a trio of Canadian paleontologists, confirmed that the teeth of ancient sharks were regularly replaced, but also made it

Photos: Dennis Fast / courtesy of the Vancouver Island Paleontology Museum

clear they have changed somewhat in shape; early sharks had cusps on their teeth, while the teeth of modern sharks have deadly serrated edges.

Getting there: The Vancouver Island Paleontology Museum is located at 587 Beach Road in Qualicum Beach. It is open from 11 a.m. to 4 p.m. Tuesday through Sunday from the May long weekend through mid-September.

Ammonoids are the most common fossils found on the eastern shore of Vancouver Island, as well as on Hornby and Denman Islands just to the east, and most are what geologists call planispirals, with their shells coiled on one plane (above). More exotic are heteromorph ammonoids (above right) that look like open corkscrews.

Below: The crab *Longusorbis cuniculosus*

266

Horne Lake Caves

RUNNING ITS LENGTH like a backbone, riddled with karst caves and sinkholes, Vancouver Island's limestone bedrock tells many tales about the island's long history. The story begins more than a billion years ago beneath the warm seas of the south Pacific. Over millions of years, as life evolved, the sandy sea bottom became a tropical reef, where corals and sponges thrived in clear shallow water. When they died, their skeletal remains, consisting mainly of calcium, settled to the sea floor, to be covered, layer upon layer, by sand and silt. Slowly, time and pressure laminated the layers and turned them to limestone.

More time passed, and the reef was thrice buried by lava during volcanic periods and gradually forced upward by undersea volcanoes until finally the submarine debris rose above the waves and became a volcanic island. Strung out, chain-like, were other smaller islands; geologists call these formations island arcs and similar formations can be seen today in places like the Philippines.

All this time, the island arc had been moving north and west on the Pacific Plate. About ninety million years ago, it collided cataclysmically with the westward moving North American plate. Vancouver Island, the Gulf Islands and Haida Gwaii had "docked" at the edge of what would one day be British Columbia.

But this is not the end of the story. In fact, as far as spelunkers are concerned, it is just the beginning. Since arriving on North America's western edge, time, the coastal rain and the island's steep topography have combined to whittle away the overlying volcanic rock, exposing in many places the soluble limestone, with its vulnerable cracks and crevices. Aided over the past two million years by waves of glaciation and the huge volumes of meltwater that inevitably followed, the ground water, with its load of carbon dioxide, has slowly dissolved the rock, ultimately riddling the island with underground passages, chambers and sinkholes.

Limestone karst caves such as these are found in other places in British Columbia, particularly in the Rockies. However, thanks to climate, accessibility, thick vegetation (which increases the level of CO_2 in the rainwater) and the island's other attractions, few are as popular as those on Vancouver Island.

Of the more than 1,000 caves on the island, a number are protected by provincial parks or recreation areas. The Nimpkish Valley, for example,

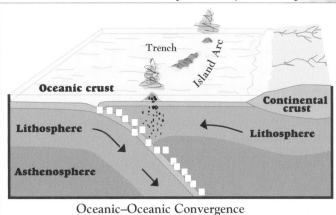

Oceanic–Oceanic Convergence

is literally riddled with caves, including those in Little Huson Cave Regional Park just west of the North Island Highway. Boasting caves featuring sinkholes, arches and a disappearing river, the park is suitable for inexperienced cavers.

One of the island's most accessible cave systems, however, is found on the Qualicum River at the northwest end of Horne Lake near the mouth of the Qualicum River. Protected by Horne Lake Caves

Main photo: Inspired by the figure in the small cave, this formation has been dubbed "The Buddha", while a guide kneels beside a formation called the "Ice Cream Flow", for obvious reasons.

Photos: Jack Most

Provincial Park, it is comprised of seven caves that boast three-storey caverns, crystal formations, cave pearls and, in the Riverbend Cave, a seven-storey waterfall.

The Pentlatch people knew about the caves along the river. For millennia, they lived along the eastern shore of Vancouver Island from the Englishman River north to Comox Harbour, as well as on Denman and Hornby Islands. The name Qualicum—which is also given to two villages, a bay and a beach, comes from *squal-li*, which in their Coast Salish dialect was the word for the treasured "chum salmon" that flowed up the Qualicum River in December or January. Low in fat, the chum were easily preserved and were therefore an important trade item for the Pentlatch people. To make the most of this trade, the Pentlatch created one of the few trails that linked Vancouver Island's east coast communities with the whale hunting Nuu-chah-nulth villages on the Pacific coast to the west.

The trail went inland along the Big Qualicum River, around the north end of a lake the people called Enoksasent (but was later renamed Horne Lake), past the caves and up over the height of land to the Alberni Inlet.

For thousands of years, people from both east and west used the trail for travelling, trading and socializing. Salmon, camus bulbs, dentalia shells, mussels and beautiful Coast Salish baskets came from the south and east, while the communities on the west traded whale oil and the magnificent furs of the sea otter and Vancouver Island marten.

Things began to change when European traders arrived in the early nineteenth century. And soon after the Hudson's Bay Company established Fort Nanaimo in 1852, the company sent twenty-four-year-old Adam Horne with the Pentlatch people on one of their trading trips to the west coast. In the ethnocentric manner of the day, the trail, the lake and ultimately the nearby caves were all soon named for a young man who travelled the region on a guided tour; for their trouble, the Pentlatch were all but destroyed by waves of small-pox that swept the island. Today, descendants of the survivors, their spouses and others are members of the Qualicum First Nation. And the Hudson's Bay Company has been sold to an American.

The caves, however, remain. Two of the most obvious, imaginatively named Main Cave and Lower Main Cave, are relatively small, respectively 136 and forty metres long, but have large galleries and narrow, but not impassable, passages. It was the

Jack Most

presence of noticeable air currents near the mouth of Main Cave that encouraged speculation about the presence of other caves in the region. Both Main and Lower Main Caves are open for self-exploration year round.

Riverbend Cave, deeper and, at 384 metres, considerably longer, has examples of rock formations that are commonly found in other

karst caves, including glistening calcite crystals, cave pearls and delicate soda straws—formed when water collects and slowly drips off a point on the roof. Stalactites, formations that grow, drop by drop when mineral-rich water drips from the roof, and stalagmites, which grow upward from the floor, sometimes join to form columns. In one area, a long series of columns has formed, creating a formation that has been dubbed The Cage.

Though it's possible that animals may have used one or more of the caves as lairs, today there are no bears, rats, bats or snakes in the caves. However, it does seem that daddy-longlegs spiders, a subtropical species unequipped for cold weather, spend the winter months inside the cave entrances.

Getting There: From the Island Highway, take Horne Lake Road about 10 km north of Qualicum Beach and, following the signs to Horne Lake Caves Provincial Park, go 12 km along a winding gravel road that traverses the scenic northeast shore of Horne Lake. (This section of the road follows the ancient east-west trail.) Cave tours of several different lengths are run daily for participants over the age of five. Though most require reservations, the 1.5-hour Family Cavern Tour is an exception. Run on a first-come, first-served basis, it begins with a 25-minute hike to the entrance and includes information on the geology of the caves, as well as a chance to see some of the formations. Caves are cool and damp, so dress warmly and wear sturdy boots or shoes. To avoid disappointment, try arriving in the morning or late afternoon. More challenging three-, four- and five-hour tours are offered on a regular basis. These, along with special tours for groups, can be arranged online. For information and reservations, go to www.hornelake.com

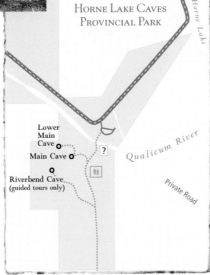

HORNE LAKE CAVES PROVINCIAL PARK

Horne Lake

Lower Main Cave

Main Cave

?

Qualicum River

Riverbend Cave (guided tours only)

Private Road

269

Courtenay &
the Puntledge River

By Heidi Henderson
and Philip Torrens

THE PUNTLEDGE RIVER gets its name from the Pentlach people, who inhabited the region for millennia before Europeans arrived. The river, like many to the north and south, is fed by snow from the mountains and by the Comox Glacier just west of Courtenay.

Despite the disastrous effects of hydroelectric generation on salmon stocks, several species still return to the river to spawn. But once the salmon runs were truly abundant: Comox (or *komoux*), after all, is a Salish word meaning "plenty".

Nor were salmon nature's only bounty. The sea provided oysters, clams, cod, halibut, and eulachon (about which more below); the land furnished cedar for houses, canoes, and even clothing, elk and deer to vary the diet, and edible and medicinal plants. Trade in many

Dennis Fast

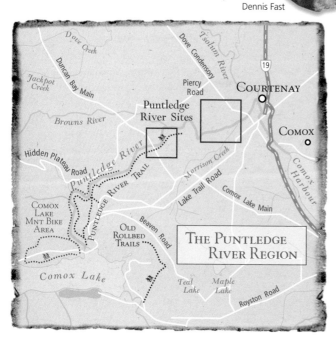

of these commodities thrived up and down the coast.

Eulachon are small fish, also known as smelt or oilfish. Like salmon, they live most of their lives in the sea, but return to rivers to spawn. It was during these returns that the Pentlach and other coastal peoples caught them in countless numbers. They are so high in oil that dried eulachon can actually be lit like wicks —hence their additional folk name of "candlefish". Dried eulachon were a valuable, portable, and compact source of fat and calories. The oil could also be squeezed out and used for seasoning, food preservation, or trade.

The same gentle climate and sheltered harbor that attracted the Pentlach continues has for decades lured both residents and visitors, particularly pleasure boaters and sport fishing tourists. More recently, additional attractions have been quite literally unearthed. The valley has become home, once again, to elasmosaurs.

In appearance, an elasmosaur resembles artists' conceptions of Nessie, the legendary monster of Loch Ness. In fact the idea that Nessie was a surviving elasmosaur was seriously proposed before one of the perpetrators of the most famous photo confessed to the hoax. Unlike the phantom of Loch Ness, however, elasmosaurs, more elegantly known

THE PUNTLEDGE
RIVER REGION

Dennis Fast

as swan lizards, were no myth.

Perhaps it is their willingness to believe in the distant past that makes children such successful "dinosaur hunters". Almost all the major finds in B.C. over the past twenty-five years have been made by children under fourteen. And so it was with Heather Trask and her father Mike who, strolling along the banks of the Puntledge River in 1988, suddenly saw the magnificent creature that would spawn a society, a museum and a new view of paleontology in British Columbia.

Elasmosaurs had never before been found in B.C. Nor had any other plesiosaurs, though similar creatures had been found on the coast of California and in central North America, where once a seaway split the continent.

Elasmosaurs swam the seas for more than 130 million years, feeding on finned fish, shellfish … and rocks. Gastroliths—round, polished stomach stones the creatures swallowed to help digest their catch, and to lower their natural buoyancy—have often been found among the bones of elasmosaurs. And in the selection of these pebbles, elasmosaurs displayed, at

times, remarkable geological discernment. The Puntledge elasmosaur, for example, had a yen for basalt. When the fossilized skeleton was excavated, the stones in its abdominal cavity were all basalt, an indication that this particular elasmosaur had somehow learned to discriminate between basalt and all other stones. Basalt, as any geologist will tell you, is harder (and therefore longer lasting as a grinding material) than many other rocks. Yet over time, even basalt will erode, particularly when subjected to the digestive juices of such an enormous creature.

From the elasmosaur's long neck and bulky body, paleontologists have concluded that it probably hunted with a sudden strike of the head, sweeping its meal into its cage-like mouth long before its "tell-tail" body loomed into sight. Swimming beneath a school of fish, and camouflaged by its dappled coloring, an elasmosaur could swing its toothy mouth up into the school, swallowing the hapless fish whole.

Less stealth was probably required to hunt ammonites—free-swimming cephalopods that jetted through the oceans like armoured squid. Distant relatives of the chambered nautilus, ammonites first appeared during the early Devonian about 250 million years ago and survived several mass extinctions before disappearing at the end of the Cretaceous. Their heavy armour likely meant that their rear view of the world was no problem when dealing with most predators, but marine reptiles were a fatal exception to the rule. Ammonite fossils with clear teeth marks have been found in many places, solving murder mysteries millions of years old.

But elasmosaurs could be victims as well as perpetrators. About eighty million years ago, a particular elasmosaur thrived in the warm, tropical seas of the deep Pacific. Paleontologists believe she lived far offshore for that was where food stocks were sufficient to sustain her great size—at more than fifteen metres long and weighing between four and five tonnes, she was exceptionally big, even by the standards of her kind. Her sisters were typically less than twelve metres, and tipped the scales at a comparatively svelte three to four tonnes.

Because their massive bodies would have made laying eggs on land quite difficult, elasmosaurs were probably viviparous, that is, they birthed their young live at sea, like whales. Modern reptiles have maternal instincts, so we can plausibly speculate that our particular elasmosaur, assuming she was in fact female, may well have invested time in teaching her young to feed.

Our elasmosaur may have perished defending her young, or in a clash of titans over a territory or perhaps in a battle for dominance. Or she may have simply died of natural causes and been opportunistically scavenged by one of her unsentimental sisters loath to forego a free meal. Regardless, she was left with the tooth of a fellow elasmosaur lodged in her throat. After death, sharks aided in dismembering the carcass; several shark teeth were found near the tip of one of her large, paddle-like limbs.

Her body, decomposing on the ocean floor, was quickly covered by mud and silt and so began the next phase of her journey—one that would turn her slowly to stone and bear her on an eighty-million year journey westward, riding the Pacific plate to Wrangellia. There she would eventually emerge beside a little river on Vancouver Island, to be found by a sharp-eyed girl and her father.

Pat Trask, below in dark sweater, doing what he does best: sharing his love of the ancient past with budding paleontologists along the beautiful Puntledge River.

Photos: Dennis Fast

Mosasaurs
by Philip Torrens and Heidi Henderson

THE PUNTLEDGE RIVER has produced bits and pieces of many other fossils since 1988. These include several mosasaurs, a group of large, long, slender, carnivorous marine reptiles that looked rather like a cross between a crocodile and a manatee. Distantly related to snakes and modern-day monitor lizards such as the Komodo dragons, they were highly successful predators. Using its flattened tail for propulsion, a mososaur captured its food and crushed it with large cone-shaped teeth. After terrorizing shallow seas worldwide for millions of years, mosasaurs disappeared in the mass extinction event at the end of the Cretaceous, along with dinosaurs and many other groups of animals.

Though large—they could grow to eighteen metres in length—mosasaurs were not dinosaurs, but rather members of the Mosasauridae family. Like elasmosaurs, they were powerful swimmers, but mosasaurs came in many different sizes and shapes, rather like today's dogs. Fossil remains from several different mosasaur genera have been found at several sites in the Comox Valley as well as on Hornby Island.

Evidence of the strength of their tremendous jaws has been found in "trace fossils"— puncture holes in ammonites and nautiloids.

Mosasaurs are also well known in Manitoba and other mid-continental sites, along the edges of an inland sea that bisected North America during Cretaceous times, but the Puntledge River mosasaurs (found in rocks that average between eighty and eight-five million years old) appear to be a subspecies, different from their prairie cousins. The first mosasaur found along the Puntledge was just 500 metres from the elasmosaur site—and soon two plesiosaur vertebrae were found nearby. Clearly, the region was a goldmine for marine reptiles, once thought to be so scarce. The best specimen was buried beneath more than twelve metres of shale and the animal appeared to be lying perpendicular to the cliff face—a lot of rock to move.

Looking uncannily like a vicious version of today's manatees, mosasaurs were not grazers but hunters, with long, dagger-like teeth, perfect for lunching on ammonites.

Photos this page: Dennis Fast, photographed at the Courtenay Museum

The Courtenay Museum

SPARKED BY THE TIRELESS enthusiasm of Pat Trask, whose twin brother Mike and niece Heather found the elasmosaur, as well as many other members of the community have who invested almost limitless time, energy and enthusiasm, an excellent museum of paleontology has been created in Courtenay. Over the years, its collection has grown to represent 400 million years on Vancouver Island.

Hanging from its ceiling, the museum has a full sized elasmosaur. Sporting a snaggletoothed grin, the

Island. But the museum is more than just fossils; it also that programs for people of all ages, as well as fossil tours that showcase the long history of the Comox Valley.

Combining fossils (of a starfish, above, and a mosasaur, below,) with interpretive murals and hands-on programming, the museum brings the past to life.

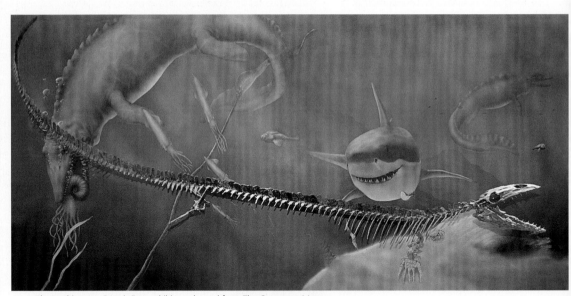

Photos this page: Dennis Fast, exhibits and mural from The Courtenay Museum

huge creature has an overbite that goes from ear to ear.

The Courtenay Museum fossil collection includes both a freshwater and a marine turtle found along the banks of the Puntledge River, a good indication that eighty million years ago, though very different in climate, the Comox Valley was a place of river mouths and seashores, much as it is today.

In addition to fossils of marine reptiles from the Puntledge, Browns, Trent and Englishman Rivers, the museum holds many invertebrate finds from Hornby Island and other well-known fossil sites on Vancouver

Getting There: The Courtenay and District Museum and Palaeontology Centre is located at 207 4th Street in Courtenay. It is open year-round from Tuesday to Saturday, between 10 a.m. and 5 p.m. For more information on tours and events, explore the museum's website at www.courtenaymuseum.ca

Bear Cove

PERCHED ON THE SOUTH EDGE of a deep narrow bay that opens onto Queen Charlotte Strait, on Vancouver Island's northeast corner, Bear Cove was sheltered from the open water. It was also buffered—at least once tides were established at or below modern levels—on the northwest and southwest by two small islands, yet it had excellent access to the bountiful waters of the strait. It was a perfect site for a village ... or a ferry terminal. In the 1970s, Bear Cove was chosen as the southern terminal for the BC Ferries' Inside Passage service.

Here, archaeologists excavating prior to the terminal construction discovered a large, multi-layered site that, at the lowest level, revealed a radiocarbon date of more than 8,000 years (which translates to a calendar date of almost 9,000 years). Situated directly across Hardy Bay from Point Hardy, Bear Cove is the oldest known human site on Vancouver Island.

Traces of the earliest occupation, which was below the many-layered shell midden, were ephemeral, but beginning about 5,300 years ago bones from northern fur seals and sea lions, as well as the remains of sea otters, porpoises, harbor seals and a small whale made it very clear that these were people who were very much at home on the open ocean. In addition to the remains of sea mammals, and bone from deer, elk and many other land mammals, in the upper layers of the site there were thousands of fish bones. All

had been preserved by the shell midden for, as lead archeologist Catherine Carlson writes, the remains of the shells "neutralized the otherwise acid forest soils". Just as important, among the bones and traces of ancient campfires, were pebble tools, cobbles of stone that had been chipped to create sharp edges, and projectile points. In the upper levels of the site, were bone harpoon points, needles, awls and chisels.

All told, the team found almost 30,000 identifiable bones; taken together, they told a story of an early campsite and later village that was regularly occupied during the winter and spring by people— very likely the ancestors of the Kwakwaka'wakw or southern Kwakiutl who still live in the region—who had mastered the complex technology of travelling and hunting on the ocean. Similar prowess has been demonstrated in ancient sites to the north, at Namu on B.C.'s central coast, which has been dated to 9,700 BP, and to the south, at the Glenrose Cannery site (see page 176) which, 8,000 years ago, was at the mouth of the Fraser River. Carlson believes that all these sites demonstrate a coastal tradition that has its origins back in the mists of time; in short, these sites were created by people of the sea, people who made "extensive use of sea fishes, sea mammals and sea birds ... all infer maritime-based adaptations."

This idea is explored more fully in one of the introductory chapters, "Peopling the Americas" on page 22.

Today, the Bear Cove site is buried beneath a parking lot at the ferry terminal, but those arriving or departing from Port Hardy can still get a sense of the place, for Hardy Bay is largely unchanged. And travelling from its sheltered confines even briefly into Queen Charlotte Strait (before ducking into the protected waters between Nigel and Balaklava Islands), allows ferry passengers some appreciation of the sea-faring skills the people of Bear Cove had developed so many thousands of years ago.

Photos: John Harvey

The Pacific Ocean shows just a little of its enormous power as it thunders against the rocks at Long Beach.

The Pacific Rim Highway

Wending its way westward

from Parksville and Qualicum Beach to Long Beach,

the centrepiece of Pacific Rim National Park's trinity of regions,

this is one of Vancouver Island's most beautiful highways.

B RANCHING WEST from either Parksville or Qualicum Beach, it travels briefly through farm country before beginning to climb the lower slopes of Mount Arrowsmith. Here, just off the highway, is Little Qualicum Falls Provincial Park, one of the most stunning public preserves on central Vancouver Island.

Divided into two sections, the park offers cascading waterfalls, clear pools, riverside trails and a camp- ground in the eastern segment, while the western section, which borders Cameron Lake (and the highway) is a popular family destination with its swimming and sailboarding. And at the west end of the park is MacMillan Provincial Park and Cathedral Grove (see page 279).

Continuing west, the highway drops into the Alberni Valley and the city of Port Alberni, which lies at the head of a long ocean inlet or fjord that more than cuts Vancouver Island in half. The fjord has long provided access to the Pacific and the bountiful Broken Group Islands (see page 285), but almost as important are the rivers that flow in to it, the Somass—originally, Tis-oma-as, "washing or cleansing"—and its tributaries, the Stamp and the Sproat. Together they constitute the most munificent salmon spawning sites on the island. Not surprisingly, both the Sproat River, which flows out of Sproat Lake, and the Somass are dotted with at least two dozen camps and village sites. In their sweeping assessment of the area, *Alberni Prehistory*, archaeologists Alan McMillan and Denis St. Claire were certain that the oldest of these sites were occupied prior to 5,000 BP.

One of the largest was the Shoemaker Bay site just north of Johnstone Island in the Somass Delta. There, though the shell midden had been consider- ably destroyed by modern construction, hundreds of artifacts were found and radiocarbon dates of more than 4,000 BP were obtained. The site was used repeatedly, and likely seasonally, over the succeeding millennia.

Home to dozens of mammals and more than 160 species of birds, the Somass and its tributaries drain

Jack Most

the much of the region adjacent
to Great Central Lake. Its estuary
is particularly important as nesting and breeding territory; recognizing this, Ducks Unlimited
and the various levels of government have embarked on the
Somass Estuary Project, to protect the intertidal marshes, mud-
flats, forested islands and lowland meadows.

    For much of the past century, Port Alberni's busi-
ness was logging, but in tune with the times, over the
past two decades it has transformed itself into a centre
of tourism, offering connections to Pacific Rim
National Park, Barkley Sound (and the Broken Group
Islands) and Clayoquot Sound. It has not forgotten its
past, however. In fact, its new focus in the past quarter-
century has shed more light on its heritage. The
Alberni Valley Museum not only celebrates the history
of the Nuu-chah-nulth peoples with excellent displays
of artifacts, but also has one of the finest collections of
basketry in the province.

    From Port Alberni, Highway 4 continues west
past Sproat Lake Provincial Park, which is known for
its magnificent petroglyph panel, K'ak'awin, at the
park's eastern end. Just a few kilometres past the long
western arm of the lake, Taylor Arm, the highway

Above: The sun sets on another day, and the ebb
tide sweeps the sands of Chesterman Beach clean,
just as they have for millions of years.

turns south and, twisting and turning,
drops off the Clayoquot Plateau.
Following the Kennedy River and the
east shore of Kennedy Lake, it heads
toward Long Beach and Pacific Rim
National Park.

Cathedral Grove

THIS TINY PRESERVE is part of MacMillan Provincial Park, which is itself only slightly larger. But within its boundaries, Cathedral Grove not only protects one of the world's most accessible stands of giant Douglas-fir but also faces a very large controversy over how to sustain this precious remnant of an ever more threatened ecology.

From the Rockies west to Vancouver Island, southern British Columbia was once covered with millions of hectares of old growth Douglas-fir (see page 281), but nowhere in what is now Canada were these remarkably adaptable trees larger or older than on southern Vancouver Island. There, thanks to

that experiences a relatively dry Mediterranean climate, Douglas-fir is the forest's "climax" species. Here, individual trees can reach heights of more than 100 metres and live more than 1,000 years.

Very little of this magnificent forest remains. In fact, this tiny preserve protects the only remaining significant grove of climax Coastal Douglas-fir. This is living history, trees that were seedlings when the fortifications at Finlayson Point were being built (see page 236) and Richard the Lion-Hearted was embarking on the Third Crusade. Here are trees that were a century old at the time of the Black Death, and becoming magnificent specimens when Columbus first sighted North America.

About 350 years ago, when these same trees were almost 500 years old, a huge fire roared through Cathedral Grove. Towering above the flames, with thick bark that was able to withstand the intense heat, the largest Douglas-firs found the blaze a blessing in

This photograph demonstrates the tremendous strength of climax Douglas-firs. While a huge winter storm—or blow-down—in 1997 toppled many of the younger specimens, the giants were largely untouched.

Dennis Fast

the island's mild climate, Douglas-fir grows from the valley bottoms to elevations of 1250 metres. And in an even more restricted area, a narrow strip along the southeastern coast

disguise, for it culled many of the smaller trees and gave the giants room to grow.

A half-century later, as they stretched silently ever skyward, their upper limbs began to sway one cold winter night. The ground shook violently beneath them and the soft, wet soil by the river collapsed; perhaps one of the titans toppled or slumped, drowning its roots and eventually spelling the end of its long life. But the damage in Cathedral Grove was nothing compared to what took place along Vancouver Island's west coast or far across the Pacific in Japan, where tsunamis washed lives and homes away.

Cathedral Grove was first thus described in the 1920s and the name fits. With its magnificent Douglas-firs, a second storey of western hemlock and western redcedar and a third storey of vine maple all soaring above, the grove is quiet and dark even on the sunniest days, with the ambience of a sanctuary.

And so it should be, for this is a place of international significance, as well as the winter range of a small herd of endangered Roosevelt elk, among many other animals. The park is also home to deer, black bears and cougars, several types of woodpeckers and owls and, in the Cameron River that flows through the park, rainbow, brown and cutthroat trout.

The park was enlarged by 140 hectares in 2005, but beyond its confines, the once awe-inspiring forests of Douglas-fir are gone, harvested for their superior-quality timber. Only one per cent of British Columbia's coastal old-growth Douglas-fir remains and, as both local and international groups have long advocated, this last remnant must not only be protected, but enlarged.

Getting There: MacMillan Provincial Park is located on both sides of Hwy No. 4, 16 km east of Port Alberni and 30 km west of Parksville. Parking is provided along the highway and, since visitors are often crossing from the north section of the park (on Cameron Lake) to the south section, where the largest Douglas-firs are found, please drive slowly. Groomed trails can be found in both sections of the park; many trails had to be recreated after a huge storm on New Year's Day 1997 toppled hundreds of trees.

Peter St. John

Dwarfed by the ancient trees on both sides, the Pacific Rim Highway (No. 4) seems more like a walking trail.

Douglas-firs

THESE MAGNIFICENT TREES, which can grow 100 metres tall and live for more than 1,000 years, are among Vancouver Island's most impressive natural wonders. After decades of logging, however, British Columbia's old growth giants are sadly few in number.

The largest of the species in Canada is the celebrated Red Creek Fir, which towers over the surrounding forest near Port Renfrew. This spectacular specimen is 4.23 metres in diameter at breast height (dbh), has a circumference of 12.55 metres, a height that was measured in 1998 to be 73.8 metres and a crown spread of twenty-three metres. The total wood volume is 349 cubic metres, which makes it the largest known individual in the entire pine family.

Pine family? one might ask. And indeed, the Red Creek Fir, like all Douglas-firs, is not a fir at all, but rather, despite its Latin name—*Pseudotsuga menziesii* or "false hemlock"—a species all its own, related to the larger *Pinacea* family. Also called red-fir, Oregon-pine, Douglas-spruce, and, in Spanish-speaking territories, *piño Oregon,* the Douglas-fir has cones that droop from the branches and fall as a whole to the ground when mature. By contrast, true firs have upright cones and scales with seeds that are shed from the cone, which remains, denuded, on the tree.

Paleontologists have determined that Douglas-firs have been a major component of the forests of western North America for the past million years, since the mid-Pleistocene. Surviving each glaciation in small refugia far to the south, these remarkable trees have marched north again at speeds that have surprised paleontologists armed with theoretical models of migration. For example, during the last glaciation, Douglas-fir found refuge in a small area at the tip of the southern Rocky Mountains in what is now New Mexico. When the world warmed, they migrated out of their refugium northwest through Utah's Great Basin at an average rate of 286 metres per year, three times faster than scientists had predicted they would.

Their magnificent wood had an even greater reach. Though the native range of Douglas-fir never extended beyond western North America, the wood was appreciated, and used, as far away as Hawaii and Polynesia, where driftwood logs from thousands of kilometres away were prized for making bowls, paddles and even canoes.

A Douglas-fir can be easily identified by recalling the four Fs. Its needles are flat, with a single groove on the upper surface; it is fragrant and friendly—soft to the touch, unlike Colorado spruce, for example—and on mature trees the bark is furrowed, with dark reddish-brown ridges. There is only one species of Douglas-fir, but two varieties— a coastal and an interior form. The coastal type is larger,

281

Peter St. John

Dennis Fast

though in most places from Vancouver Island to northern California, it will eventually give way to western hemlock as the climax or dominant species. In a narrow, drier band on Vancouver Island's southeast coast, however, Douglas-fir is the climax species; Cathedral Grove is a superb example of such a climax forest.

The interior variety of the species is comparatively tall and slender, and grows in a great sweep east to the Rockies and south along the western mountains all the way to Mexico.

Native British Columbians used the dense, durable wood of the Douglas-fir for a myriad of things, including bowls, spears, harpoon hafts and barbs, fish hooks and traps, snowshoes and burial caskets, as well as fuel for longhouse firepits. The boughs were used as floor coverings; the seeds were eaten and the sap was used for calking canoes. Sap was also used to saturate the tips of heartwood pegs to use as torches that would burn even in the rain and as a medicinal salve for small cuts and sores.

As the "frosting" on this remarkable tree's contributions, each summer branches in sunny locations exuded something that has been called "Douglas-fir sugar", which was much prized by early North Americans, who lived without honey or cane sugar. Tests have shown that the frosty substance is fifty per cent trisaccharide sugar.

Insects, birds and mammals of many species also rely on Douglas-firs for food and shelter.

Canada's oldest known Douglas-fir was one of a stand on Vancouver Island that was established after a fire about 635 AD. It blew down in a storm in 1985 and was afterward reported to have a ring-counted age of 1,350. Other trees are still alive in the same stand.

The oldest accurately-dated Rocky Mountain Douglas-fir is a 1,275-year-old giant in New Mexico. This longevity is apparently an anomaly; growing on a relatively barren lava field has protected the ancient tree from fire, animals and humans. The oldest authenticated Douglas-fir ever found in the U.S. was slightly older than 1,400 years of age when it was harvested in the 1950s near Mount Vernon, WA.

Historical records document trees over 120 metres tall, though specimens over eighty metres are

DAVID DOUGLAS
Scottish Botanist

Douglas-firs were named for David Douglas, a Scottish botanist whose collecting trips were financed by the Horticultural Society of London. He arrived in British Columbia in 1825 at the age of twenty-five. Travelling with a small terrier, living off the land, sleeping under his canoe or rolled in a blanket, he trekked thousands of kilometres through the province. Following his adventures in B.C., Douglas returned to England, bearing seeds from a Douglas-fir. These were planted at Scone Palace, as part of the collection of Kew Gardens, and there one survivor remains, one of the tallest trees on the grounds today. Douglas died a violent and mysterious death in Hawaii in 1835. He was found dead in the bottom of a pit-trap created to capture wild bullocks. One of these huge creatures, also dead, was in the pit with him, but Douglas's wounds did not seem consistent with those he might have received from the bull. Some believed he had been murdered, but the cause of his death was never determined.

very rare today. In Washington state, the tallest tree on record was found near Little Rock. It was 100.5 metres tall, but less than two metres in diameter.

Clayoquot Sound
UNESCO Bioshere Reserve

FOR MOST PEOPLE, Clayoquot Sound will be forever associated with the largest environmental protest in Canadian history. Beginning with concerns about MacMillan Bloedel's plans to clearcut the forests on Meares Island, directly north of Tofino, in the 1980s, the protest escalated into a blockade. Then, as tensions built between the Nuu-chah-nulth and non-native residents of the region on one side, and the logging company on the other, it grew into a massive public outcry during the summer of 1993 over the provincial government's plans to protect very little of Clayoquot Sound's magnificent coastal temperate rainforest. That summer, as 856 people were arrested for their role in protests and blockades and world attention focused on the campaign, a revolution in forestry practices was born, including a movement from clearcut to ecosystem-based logging. Seven years later, Clayoquot Sound was designated a United Nations biosphere reserve— in essence a living laboratory for sustainable development.

It's a concept with which the Nuu-chah-nulth people have long been familiar; they call it *hishuk-ish ts'awak*: "everything is one". In other words, impacting one aspect of life —be it cultural, environmental or economic—affects everything else.

The nations of the sound have learned this over many millennia. Moving to take advantage of seasonal resources, over time they built handsome villages on many of the islands, both large and small. The forests that ultimately spurred the protests

provided homes and transportation on a scale early Europeans found hard to describe.

"On entering the house we were absolutely astonished at the vast area it enclosed," wrote Captain John Meares on arriving at the spring village of the Tla-o-qui-aht on the island of Echachist in 1788. "It contained a large square, boarded up close on all sides to the height of twenty feet, with planks of an uncommon breadth and length. Three enormous trees, carved and painted, formed the rafters, which were supported at the ends and in the middle by gigantic images, carved out of huge blocks of timber."

Yet this village, located on one of the outer islands of the sound and apparently used to hunt gray whales

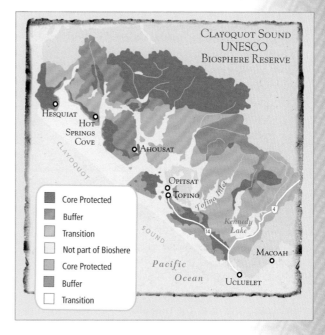

CLAYOQUOT SOUND
UNESCO
BIOSHERE RESERVE

HESQUIAT
HOT SPRINGS COVE
AHOUSAT
OPITSAT
TOFINO
Tofino Inlet
CLAYOQUOT
SOUND
Kennedy Lake
4
14
Pacific Ocean
MACOAH
UCLUELET

- ■ Core Protected
- ■ Buffer
- ■ Transition
- □ Not part of Biosphere
- ■ Core Protected
- ■ Buffer
- □ Transition

during their spring migration up the coast, was small when compared to the summer and winter homes of the Tla-o-qui-aht at Opitsat, directly north of today's Tofino, or at Clayoquot, just to the west.

Clearly, old-growth trees, whether cedar or Sitka spruce, which reached enormous sizes on the islands and in the valleys east of the sound, had been selectively used for thousands of years by the people who

Jack Most

lived here. Despite this, to European eyes, the forests seemed untouched, "a wilderness". It is these forestry practices—the careful selection, diacritical logging and eye to the overall health of the forest practiced by their ancestors—that native logging companies, established since the reserve was created, hope to emulate.

To see the difference between a forest that has served people for millennia without obvious harm, and one that has been indiscriminately cut and destroyed in less than a century, the Western Canada Wilderness Committee, with the blessing of the Tla-o-qui-aht people, have created the Clayoquot Valley Witness Trail, a challenging fifty-eight kilometre return trip that takes hikers through clear-cut areas (around the north and south trailheads) and ancient rainforest in-between.

To see some of the islands in Clayoquot Sound, tour operators offer day trips from Tofino to several of the nearby islands, or passage on the water taxi to Meares or Vargas Islands. Guided tours can also be booked with the Nuu-chah-nulth Booking and InfoCentre in Tofino.

A little farther afield, the community of Ahousaht on Flores Island is building the Ahousaht Wild Side Heritage Trail, a rustic and moderately strenuous ten kilometre trail leads from their ancient village along the south side of the island to Cow Bay in Flores Island Provincial Park. Cutting across rocky headlands and sandstone reefs, the trail also winds through old-growth forests of Sitka spruce and allows access to some of the island's beautiful white sand beaches. In the spring, Flores Island is an excellent place to see gray whales in migration between March and May, but keen-eyed observers may even see some of the resident grays that use the bays for feeding during the summer months. Flores is twenty kilometres north of Tofino and accessible by water taxi, kayak or floatplane.

Across a narrow channel from the island's west side is the very popular Hot Springs Cove, felt by some to be the finest hot springs in Canada. Accessed from the government dock (which is a regular stop for water taxis and float planes from Tofino) by a 1.5-kilometre cedar boardwalk through old-growth forest, the springs bubble from the bedrock in a stream that's 51°C—too hot for comfort—and flow over a small cliff into the tidal pools below. The temperature of these soaking pools depends on the tides, which also ensure that the area is flushed clean twice daily.

Getting there: To access the Clayoquot Valley Witness Trail, which is best hiked from south to north, turn north off Hwy 4 onto the Kenquot Main Logging Road 11.5 km west of Sutton Pass. The gate to the road is normally locked, so park and hike in from here. The trailhead is about 7 km beyond the gate. There are no services along the trail and fresh water may not be available during the summer. For a trail map and complete directions, contact the Western Canada Wilderness Committee in Vancouver at 1-800-661-9453.

As indicated, there are a variety of ways to reach the various islands in the sound. To inquire about options, call the Nuu-chah-nulth Booking and Info-Centre at 1-800-665-9425.

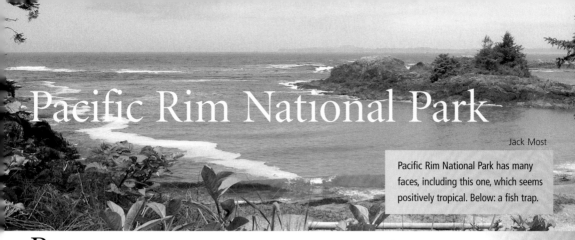

Pacific Rim National Park

Jack Most

Pacific Rim National Park Reserve is many things to many people. In part, that's because it represents three, very different ecosystems stretched along the southern half of Vancouver Island's magnificent west coast. Long Beach, the most northerly, faces the Pacific with bays and coves and apparently endless sandy beaches between today's communities of Ucluelet and Tofino. South of Long Beach, the Broken Group Islands, a scattering of more than 100 islands, islets and rocky outcrops is centred in Barkley Sound. And south again is the West Coast Trail, a rugged seventy-five kilometre stretch of coastline dominated on the landward side by temperate old-growth forests of spruce, hemlock and cedar, and seaward by a treacherous shoreline well known as the "graveyard of the Pacific".

For visitors today, Long Beach is by far the most popular segment of the park, though the Broken Islands draw increasing numbers of paddlers and the West Coast Trail is almost always fully booked with determined hikers during the summer season. But this otherworldly coast is more than just a place to visit. For today's fourteen Nuu-chah-nulth nations, and their ancestors, the land "all along the mountains" has been home for

millennia. This is why Pacific Rim is a "national park reserve"; it sits on land that has been claimed by the Nuu-chah-nulth and Pacheenaht people. Backing that claim is a wealth of archaeological sites that lie within Pacific Rim's boundaries.

The park's own information cites a total of 290 archaeological sites that have thus far been identified; these include shell middens from villages—some of them established for thousands of years—camps and defensive locations; fishing and burial sites; petroglyph and pictograph sites and "CMTs" (see page 288), culturally modified trees that have been used in a variety of ways by the people of the region. The population in the park's three units has been estimated at about 12,000 in the late 1700s or early 1800s, when disease had already begun to take a toll.

Surprisingly, perhaps, given the popularity of Long Beach today, the large majority of the villages and camps were found not there, but on the Broken Group Islands. Here, on even the most exposed islands, villages or camps that were used at least seasonally were established well before Hammurabi became king of Mesopotamia. One of the oldest settlements, Ts'ishaa, was located on what is now known as Benson Island, which directly faces the Pacific on its northwestern side. Here, over a period of three years, Vancouver archaeologists Alan McMillan and Denis St. Claire excavated not just one, but two villages on the site, occupied consecutively over a period of more than 5,000 years, making it the oldest archaeological site on Vancouver Island's west coast. McMillan and St. Claire were ably assisted by members of today's Tseshaht community, whose ancestors lived at Ts'ishaa—in fact, Tseshaht means, literally, "the people of Ts'ishaa"—but make their home today at Port Alberni.

Peter St. John

Opitsat today, with Lone Cone Mountain rising behind it. The ancient community is directly north of Tofino, on Meares Island.

The excavations, which began on the main village site, quickly uncovered deposits of clam and mussel shells up to four metres deep. Clearly, these were people whose living came from the sea. But the Tseshaht were also big game hunters; large stacks of whalebone, including one whale skull with an ancient harpoon head of mussel shell deeply embedded in it, were testament to their remarkable skills. Ts'ishaa, by the way, means "where the whalebones on the beach smell".

In total, more than 700 artifacts were recovered from the village, not only including various pieces of fishing gear made of bone, but also a decorated bone comb and a beautifully polished jet pendant, demonstrating that life was not all work and no play.

The village midden dated back about 2,000 years, but on an elevated ridge behind the village, the team discovered a much earlier site. First occupied when sea levels were about three metres higher than at present, radiocarbon dating showed the original village to be more than 5,000 years old. "Waves once broke at the base of this ridge," McMillan explained in an account of the project, "prior to the gradual lifting of the land and the buildup of the later village below."

Artifacts excavated from the older site were mainly of stone and rather crudely manufactured, quite different than the sophisticated objects found in the later settlement below.

Sprinkled throughout the islands, at least eighty other villages and camps have been found, as well as dozens of oceanside stone fish traps, a clear demonstration that the Nuu-chah-nulth people not only populated the islands of the Broken Group, but used all its resources. And as expected, given the number of settlements, the vast majority of burial sites thus far discovered have also been found on the islands.

However, both the Long Beach and West Coast Trail regions of the park were also quite densely populated prior to the arrival of Europeans, or the diseases that often heralded their coming. More than thirty villages and camps have been found in the Long Beach portion of the park, and forty have been located between the modern communities of Port Renfrew and Bamfield. In addition, most of the petroglyphs discovered in the region, as well as indications of trees that have been used, are along the West Coast Trail. Given the trail's proximity to the magnificent forests of the Carmanah and Walbran Valleys, which even outsiders find inspiring, this should come as no surprise.

Much has yet to be learned about the villages of the region, but two stand out. One, just outside Pacific Rim National Park Reserve at Little Beach in today's Ucluelet, was first occupied about 4,000 years ago. Most of the artifacts recovered were of stone, which had been pecked, ground or chipped, rather uncommon in a region known for its artifacts of bone, shell and antler. Among the stone points was one that was large and leaf-shaped, very unusual on the West Coast. The whole collection was reminiscent of the Shoemaker Bay site at Port Alberni, a village that was known to have been occupied by Salish-speaking people prior to their "being forcibly assimilated into Nuu-chah-nulth culture in the early 19th century", according to archaeologist

Jack Most

Richard Brolly. Was the Little Beach site a Salish outpost as well?

The other village sat on Pachena Bay, near the north end of what is now the West Coast Trail. At 9 p.m. on the night of January 26, 1700, it—like many other villages across the southern half of Vancouver Island—began to shake violently. People who had not yet gone to bed likely found it hard to stand. The timbers of the longhouses began to collapse and outside, trees in the forest crashed to the ground as rocks on the cliffs along the shore tumbled into the sea.

One can imagine the people stumbling outside as the beach rose and fell beneath their feet, for elsewhere, the shaking was so prolonged it made people ill. When at last it stopped, there must have been confusion and fear among the people of Pachena Bay. What evil spirits were about? Was it the end of the world?

It's unlikely that any of them saw the great wave as it rose from the dark ocean. Thundering onto the beach, it completely destroyed the village, smashing the longhouses and sweeping the people out to sea. When

Getting there: The various sections of Pacific Rim National Park Reserve must be reached in various ways. Long Beach is located along Hwy 4, 108 km west of Port Alberni. The highway also leads to both Tofino and Ucluelet. The latter, which is just south of the park, can also be reached by boat from Port Alberni during the spring, summer and fall. Wickaninnish Interpretive Centre, located in the park and open from mid-March to mid-October, focuses on the area's natural and cultural heritage. The Green Point Campground, which is usually full during the summer months, takes reservations for drive-in sites and the group site.

Canoeists and kayakers can access the Broken Group Islands via passage on the *MV Lady Rose* and the *MV Frances Barkley*, which travel regularly between Port Alberni and Sechart, on the edge of the archipelago.

Those interested in hiking the West Coast Trail must register well ahead. Parks Canada's website has full information on all three park units.

neighbors came to investigate in the days that followed, the village had vanished. And there were no survivors.

The violent shaking was the result of one of the largest earthquakes in recorded history. Estimated to have been a magnitude 9 on the Richter Scale, it occurred along an undersea fault that stretched 1,000 kilometres down North America's west coast, from the middle of Vancouver Island to northern California. It produced a tsunami that not only destroyed the village at Pachena Bay, but raced across the Pacific to cause destruction and death along the coast of Japan.

Though well educated about the hazards of earthquakes, for generations the Japanese could not explain the tsunami of January 1700. At last, almost 300 years later, using dendrochronology, tectonic data and oral histories, scientists in the U.S., Japan and Canada were able to solve the mystery.

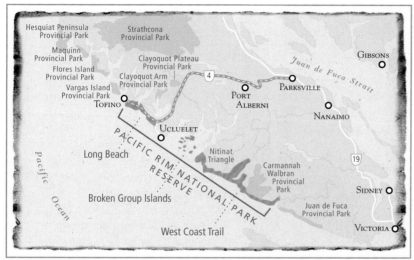

The Living Forest

NOTHING DISTINGUISHES native North Americans from those who have arrived in the past 500 years more than their approach to nature. No matter where they lived, the continent's original inhabitants almost invariably viewed themselves as part of the natural world. By contrast, settlers from beyond North America's shores saw themselves as separate from nature and, perhaps more important, viewed nature's once breathtaking bounty as apparently endless, riches to be possessed and utilized without reserve or thought for the future. Today, the result is evident everywhere in seas that are all but empty of life, mountains denuded of trees, and overgrazed and infertile grasslands.

Despite centuries of determined pillaging, Canada's enormous size and magnificent diversity

Bark from redcedars was taken in long strips that had been cut at the bottom. The tree was then left to heal.

combined to protect pockets of nature's bounty. And thanks to its steadfast inhabitants, its challenging geography, an increasingly vocal environmental lobby and, in recent years, the reluctant support of its provincial government, many of these preserves are in British Columbia. From Gwaii Haanas and the Great Bear Rainforest to Clayoquot Sound and Cathedral Grove, they allow us glimpses of the magnificent world of the past.

Many of us call these old-growth reserves "wilderness", assuming that forests with trees 500, 800 or 1,000 years or older are somehow wild, and therefore "untouched by human hands". As Jeffrey McNeely wrote in 2004, "The idea that nature exists in isolation from people has become part of the mythology of industrial society." Yet nothing could be farther from the truth. In fact, as the aboriginal inhabitants of many parts of the world have always known and as researchers are beginning to discover, old-growth forests have always been a crucial part of the lives of many cultures.

The Haida, the Heiltsuk and the Nuu-chah-nulth, to name but a few northwest coastal peoples, used more than twenty species of trees for almost every aspect of life. Most important was western redcedar, the "tree of life" for many coastal cultures. In a summary report for the David Suzuki Foundation, Arnoud Stryd and Vicki Feddema recently wrote that this magnificent tree, which can grow more than 100 metres tall and live more than a thousand years, provided

Amanda Dow, both pages

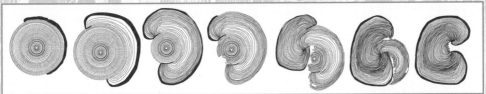

Dawn Huck, after illustrations by Millennia Research

Beginning at left, these illustrations show how trees from which bark has been culled heal themselves. A) Bark is stripped in 1799; B & C) scars form as tree grows; D) bark is stripped again in 1817; E-G) first scar lobe grows into original tree, while second scar face decays.

bark for clothing, hats, mats, twine, blankets, diapers, towels and rope. It furnished planks for longhouses, logs for canoes, branches for fish traps, roots for baskets and cradles, and wood for masks and totem poles, art that has come to symbolize both British Columbia and Canada.

For at least the past 5,000 years, dozens of cultures have extensively used western redcedar, and many other species of trees have been used for 10,000 years or more. Yet when Europeans arrived, they perceived the forests to be "untouched".

These apparent contradictions can be quite simply explained. B.C.'s coastal cultures harvested what they needed of a given tree and left it to heal itself. If, as in the case of the creation of one of the magnificent Haida sea canoes or a Nisga'a or Kwakwaka'wakw totem pole, an entire tree was needed, it was taken only after the appropriate rituals and prayers were performed. Clear-cutting was not only impossible because of the available tools, it also was unthinkable. Destroy the tree of life and life itself would be destroyed.

Only recently have archaeologists begun to understand this cultural process and its place in reconstructing the past. Not surprisingly, perhaps, they have created their own label for the trees that have been thus marked by generations of judicious use— culturally modified trees, or CMTs. As Stryd and Feddema write, "For the Heiltsuk First Nation near Bella Bella, resource gathering sites (including CMT sites) are of spiritual significance because the resources were believed to be a gift from the Creator."

Such areas are also often an indi-cation of long-term use, even where —as in Stanley Park or Cathedral Grove—the original inhabitants no longer reside. Because trees were used in different ways, at different times, for different purposes over centuries, and because of their remarkable ability to heal themselves, detecting the trees thus used—the CMTs—is not always easy.

Nor are CMTs restricted to the forests of North America. Recent research by scientists in other parts of the world has revealed that living trees were also used in northern Scandinavia and southeastern Australia, among other places.

In his 2005 doctoral thesis at the Swedish University of Agricultural Sciences, Rikard Andersson recorded that in northern Sweden, the Sami, a rein-deer herding society, used different tree species for a variety of reasons. Silver birches were scarred down to the cambium on one side of the trunk to produce— after twenty to forty years—a thickening growth along the scar that made remarkably strong axe handles. Large Scots pines were sometimes carved near the base of the tree to produce handles for fastening rein-deer cows during milking, or to attach a fence of some kind. During times of great scarcity, the bark of pine trees was also peeled so that the inner bark could be eaten. This practice apparently goes back at least 2,800 years, according to C–14 dating of subfossil logs.

In Australia, the Aborigines used bark from living red gum and gray box trees as roofing material, as well as to make canoes and to fashion shields and contain-ers. Research into CMTs is ongoing in other parts of the world, including the boreal forests of Russia and the mountain regions of the southwestern U.S.

Visitors to many old growth forest reserves that have been recently protected—Cathedral Grove or even Stanley Park—may see trees with scars that record their part in an integrated circle of life. It's an intensely spiritual experience that should make every-one think twice about today's disposable world.

The MV *Uchuck III* threads through the islands of Resolution Sound. The converted mine sweeper provides access to Yuquot and area villages.

The Gold River Highway

Stretching from Campbell River

on the Strait of Georgia to Gold River, just north of Muchalat
Inlet, the Gold River Highway (No. 28) provides a dramatically
beautiful link between Vancouver Island's east and west coasts.

IT ALSO CONNECTS the traditional territories of the Kwakwaka'wakw people of the northeast coast with the Nuu-chah-nulth of the Pacific, all in less than 100 kilometres.

This is not a highway to rush, however, for the island's rugged natural beauty is everywhere, beginning just three kilometres north of Campbell River at Elk Falls Provincial Park. Here, trails lead through mature second-growth Douglas-fir forests to canyons, the banks of the lower Campbell River and thundering Elk Falls, for which the park is named. For avid fishers, locals boast that this is the "finest year-round salmon fishing in the world".

From the park, Highway 28 heads west, circumvents the southern end of upper Campbell Lake to bisect Strathcona Provincial Park (see page 293), the oldest provincial park in British Columbia and the largest on

Vancouver Island. Gold River is just outside the park's western boundary.

Despite its name, which recalls the gold that was taken out of the area in the 1860s, Gold River was a created a century later, in the mid-1960s, and owes its existence to the pulp and paper industry. It's a thoroughly modern town; in fact, it was Canada's first all-electric town and the first to boast underground wiring.

Today, however, Gold River's focus is on the increasing appreciation for the island's long history and wild beauty. Visitors heading for Nootka Sound (see page 295) and Yuquot, its 4,000-year-old centre (see page 297), leave from Gold River on the *Uchuck III*, a converted minesweeper that docks at the head of the Muchalat Inlet fourteen kilometres southwest of town.

Cavers are drawn by Gold River's reputation as the "caving capital of Canada" and particularly by the Upana Caves, a series of caves formed in the limestone bedrock by an underground stream. Located just seventeen kilometres west of Gold River on Head Bay Forest Road, the caves have at least fifteen entrances linked by trails; the most

Jack Most

291

impressive is the large lower stream entrance.

Visitors should bring helmets, lights, warm clothing—some of the caves are the deepest north of Mexico—and a camera. Parking and washrooms are available on site. Both self-guided and guided tours are available.

Gold River was also known for a time as the home of a young orca or killer whale that turned up in Nootka Sound in July 2001, after being separated from L-Pod far to the south in Washington State's Puget Sound. Early attempts to reunite the young whale with his family were abandoned at the request of the local Mowachaht and Muchalaht people, who believed Tsu'xiit (also known as L–98 or Luna) embodied the spirit of Chief Ambrose Maquinna, who had said he would return to his people as a whale as he died just days before the year-old orca turned up.

For nearly five years, the whale frolicked in the inlet, rubbing against boats, entertaining visitors and occasionally getting in the way of float planes or annoying fishermen. Then in March 2006, despite considerable efforts on the part of the Nuu-chah-nulth people to educate boaters, the six-year-old whale got too close to a tugboat propeller and was killed. A movie, *Luna: The Way Home*, starring Adam Beach and Graham Greene, has been made of the life of a remarkable animal.

Tahsis Inlet stretches north from Nootka Sound to Tahsis, the modern headquarters of the Mowachaht and Muchalaht First Nation.

N.T. Burroughs, BC Archives , PDP-9054

Strathcona Provincial Park

Jack Most

Huge and triangular in shape, Strathcona Provincial Park is a jewel that stretches almost coast to coast on Vancouver Island's midriff. Encompassing elevations from sea level, at Shelter and Herbert Inlets off Clayoquot Sound, to the 2,200-metre summit of the Golden Hinde, the island's highest point, it includes more than 250,000 hectares. Within its boundaries are old-growth fir and cedar rainforests, extensive areas of karst with caves and sinkholes, six of the island's seven highest mountains, its only remaining glacier, a plethora of lakes and rivers and one of the ten tallest waterfalls in the world.

This wealth of scenic wonders didn't come about all at once, of course. In fact, the Strathcona Park we know today was almost 400 million years in the making and born half a world away.

Like many other parts of what geologists call the Insular Super-terrane, the central part of Vancouver Island began as an ancient series of volcanic eruptions on the floor of the ocean far to the east, about where modern Indonesia sits today. Over a period of about 200 million years, three distinct periods of volcanic activity created a long, curving arc of undersea mountains that eventually breasted the waves to become islands.

At times the volcanoes erupted violently, with surging, exploding lava and massive rock slides; at other times, the eruptions were quiescent, oozing lava in great sheets. And within and between these volcanic periods, when the mountains were quiet, beneath the water, great beds of marine corals and ancient sea life grew on the rock, and sand sifted down, creating what would one day be limestones and sandstones. The limestone, with its fossilized stems of crinoids—or sea lilies—which, since they are often seen in cross-section, are shaped like tiny doughnuts, can be best seen in the Buttle Lake formation at Marble Meadows area and along Karst Creek These fossils have allowed paleontologists to date the Buttle Lake formation to between 290 and 260 million years ago.

The second volcanic period produced what geologists call the volcanic flood. Lava flowed out of long fractures in the sea bed and, chilled by contact with the water, formed pillow-shaped blankets that ultimately piled more than 6000 metres thick. This is the rock that constitutes most of the bedrock in the park. It can be seen in road cuts along the mine road from about 1.5 kilometres south of the park headquarters to beyond Lupin Falls.

When the flood ended, layers of fine limestone and limey shale slowly built up. About 190 million years ago, they were topped by a final cycle of volcanic activity that at last breasted the waves to form a long chain of volcanic islands that would one day be known as Wrangellia.

Over millions of years, the island arc moved east on the Pacific plate, occasionally colliding with other land masses or terranes that crumpled and folded the mountains. About ninety million years ago, the island chain cemented itself into the westward trending edge of North America. Wrangellia now stretched north from southern Vancouver Island through Haida Gwaii and the Yukon into southern Alaska's Wrangell Mountains. The building process was at last complete.

The last sixty million years have seen the final sculpting: the weathering of the rock by rain and wind, the smoothing of the peaks and carving of the valleys by waves of glaciation, which also deposited mud, sand, gravel and boulders in the lowlands and

Above: Mount Myra, named for Lord Strathcona's daughter, soars skyward.

Getting There: Strathcona Park is located on central Vancouver Island. The Gold River Highway (No. 28) from Campbell River runs through the northern part of the park, while Westmin Road branches south from 28 into the heart of the park.

valleys, and left deep scratch marks—glacial striations—on rocky ledges. As the landscape changed, so did the animals and, ultimately, the humans who called it home. Its wealth of ecosystems, gave Strathcona Park a great variety of plant and animal species, and that in turn, particularly over the last 10,000 years, made it a prime destination for hunters, fishers and trappers. As time passed, the people settled mostly on the coast, and became marvellous whale hunters, at home on the open ocean. But they did not forget the bounty of Strathcona Park and returned there every fall to hunt deer and elk and, as we have recently discovered, marmots. Yet, perhaps in part because of its challenging topography, very little is known about how or where people lived in this region.

A tiny window on the past has opened over the past two decades with the discovery, usually by climbers or cavers, of piles of Vancouver Island marmot bones in two caves in Strathcona Provincial Park, as well as others in a karst cave in Clayoquot Plateau Provincial Park east of Pacific Rim National Park and a small granite rock shelter in Limestone Mountain south of the Alberni Inlet.

Under the auspices of BC Parks, specialists have visited all the caves, examined the sites and taken samplings of the bones. The resulting radio-carbon dates ranging from 2,600 to 800 years ago, along with cut marks on many bones, showed not only that marmots had been hunted for pelts and for food, but that they were numerous in sites from which they are extirpated today.

As later fur traders on the coast, who collected marmot pelts in huge numbers annually, discovered, the fur of the little grazer is light and warm, perfect for blankets, cloaks and capes, while the meat is—according to historic sources—very tasty.

Finding bones from marmots of all ages—young-of-the-year, young adults and older adults—made several things clear; the animals were almost certainly taken in the immediate areas of the caves, very likely in September or early October, after their winter fur had grown in but before they went into hibernation in late October.

During the period in question—between 2,600 and 800 BP—the Nuu-chah-nulth people lived in many villages along the Pacific Coast, both on the mainland and the many offshore islands. Though most of their livelihood came from the sea, they hunted inland as well, usually in September and October, when the elk and deer were fat and their hides were in prime condition.

The discovery of caches of marmot bones indicates that marmots were also hunted during these months, though whether on a regular basis—as was the case much later during Vancouver Island's relatively brief fur trade period—or only at specific times, for the creation of magnificent marmot capes to be presented at potlaches and other festivities, is not known.

In addition to archaeological information, the bones also told biologists a good deal about the original ranges of the now grievously endangered Vancouver Island marmot, information that might be used to repatriate the animals—which are now being successfully bred in captivity—to areas where they once thrived.

Myra Falls cascades into Buttle Lake

Jack Most:

Nootka Sound

Beautiful Yuquot (or Friendly Cove as eighteenth-century English sailors dubbed it) is today a National Historic Site.

Jack Most

THOUGH MAGNIFICENT red-cedars and yellow cedars towered over the shores, though the forests were filled with deer, elk and bears, like the people who lived around many of the world's other great harbors, for thousands of years the lives of the people of Nootka Sound depended mainly on the sea. From Yuquot (see page 297) to Tahsis, the "gateway" to the grease trail across Vancouver Island, the sound provided everything the Mowachaht needed to create a culture that was rich, bountiful and lasting.

For millennia their lives centred on whaling. Armed with an intimate knowledge of the ocean, an almost spiritual connection with its inhabitants and long spears tipped with points of mussel shell, they hunted gray and humpback whales far out on the Pacific (see page 299), as well as seals, sea lions and porpoises closer to shore. This prowess brought great prestige, for only the Mowachaht, their Nuu-chah-nulth and Pacheenaht neighbors, as well as the Makah of the Olympic Peninsula hunted great whales on the open ocean.

But Nootka Sound offered much more than whaling: halibut and cod in the waters offshore; rivers where eulachon (rare on Vancouver Island) and five species of salmon spawned; tidal shallows full of shellfish and shores where berries of many kinds, as well as bulbs and tubers grew in abundance.

Over the millennia, these riches allowed the people to create a society of demonstrable wealth. Great long-houses of cedar lined the coves and beaches; inside, the posts and beams were magnificently carved. Little wonder when European ships began arriving at the end of the eighteenth century, the artists on their expeditions were so struck by the extraordinary art and design—which may have seemed baronial to their eyes—and the beauty of the surroundings, that they couldn't stop painting. As a result, a remarkable number of the early paintings of British Columbia are of Nootka Sound and Yuquot (or, as the English named it for the reception they got on their unheralded arrival, Friendly Cove).

Despite this apparent wealth and their prowess at sea, the Mowachaht were not in a position to take their situation for granted, for they had enemies to the south around Clayoquot Sound, and to the east, where the Muchalaht lived along a long inlet that extends far into Vancouver Island. (Today, these eastern neighbors have joined with their former rivals to create the Mowachaht/Muchalaht First Nation.) Border disputes and skirmishes seem to have been a regular part of life. Just months after Captain James Cook and his English ships arrived at Nootka Sound in late March 1778, the chief or Maquinna of the Mowachaht was killed in a battle. (Cook, as it turned out, did not have

295

National Archives of Canada C–003676

Jack Most

Above: John Webber drew this illustration of a house at Yuquot in 1778. Below: Facing the open ocean, the outer beach at Yuquot is five-minutes from the dock.

fur of any mammal on Earth. By 1793, at least 100,000 pelts had been taken from Nootka Sound, according to the ships' logs of the time, which were famous for under-reporting their take. By 1803, the trade had moved north to Haida Gwaii and within a century, sea otters had been extirpated from the waters off British Columbia.

Finally, as the world population hung in the balance, these charismatic animals were at last protected by law. Between 1969 and 1972, eighty-nine sea otters were transplanted from one of the world's last remaining populations in Alaska to the coastal waters of B.C. and slowly their numbers began to climb. Today their population off Vancouver Island is estimated to be about 2,000 and once again sea otters can be seen in Nootka Sound.

long to live either. He was killed the following year in Hawaii.)

The Maquinna was succeeded by his son, who must have been little more than twenty at the time, for English fur trader John Meares, visiting Yuquot in 1788, wrote that he "appeared to be about thirty years". Yet this young man assumed more than his father's title; he also displayed wisdom beyond his years. Over the ensuing decade, he dealt with the Spanish, who established a small trading post—Santa Cruz de Nutka—at Yuquot, and with dignity and foresight met the ever-greater number of ships that thronged to the sound to get a piece of the almost unbelievably lucrative sea otter fur trade. This enlarged Maquinna's reputation, not only among the visiting Europeans, but also among his rivals beyond Nootka Sound. Today, one of his descendants, Michael Maquinna, is chief of the first nation.

However, the fur trade also had ruinous effects. European contact brought disease, death, and destruction of many of the sound's once apparently limitless natural resources. The first to be devastated were the sea otters. Cute and cuddly, sea otters have the densest

Getting there: A number of sites in Nootka Sound, including Yuquot, Bligh Island and Tahsis, the modern headquarters of the Mowachaht/Muchalaht people, are regularly accessed by the MV *Uchuck III*. Leaving Gold River, at the end of the Gold River Highway (No. 28) on a regularly scheduled basis, the converted minesweeper takes up to 100 passengers and up to 100 tonnes of freight. Informal stops can be made to wet launch kayakers and ocean canoe parties, and arrangements can be made for side trips to off-lying islands.

Yuquot

HERE, ON CANADA'S far western
edge, is a place that two peoples,
both fundamental to what British
Columbia is today, count as a "birthplace". In the Wakashan language
of the Mowachaht people, who have
lived here for at least 4,200 years,
that place is Yuquot, "where the wind
blows from all directions". This is
where their culture was born. Here,
in the forests by the sea, the first
Mowachaht man was born, say the
legends of the people, formed of the
tears and mucous of the first woman.
Their offspring ultimately peopled
dozens of villages around the sound.
The largest was Yuquot, which eventually had more than 1,500 residents.

The English, who arrived in
the late eighteenth century, and the
Spanish who followed them—in the
Mowachaht language, both were
mamalthi, "people who live on a ship"
—had another name for Yuquot.
They called it Friendly Cove, for it
was here that they found shelter
from the storms, and here that they
were met with aid and cooperation.
For the Mowachaht, that cooperation
ultimately had a very high price. for
as the decades passed it led to hard-
ship, disease and death.

It's not surprising, perhaps, that
the meeting of such contrasting cultures would lead to misunderstandings.
Even the name of one of the world's
great harbors recalls the confusion

John Webber, National Archives of Canada C-6641

John Webber, who travelled with Captain Cook,
recorded this meeting of two disparate cultures.

Getting there: Yuquot is accessed by the MV *Uchuck III,* a
converted minesweeper that regularly plies the waters of
Nootka Sound from Gold River. Travelling to Yuquot on
Wednesdays and Saturdays, the schedule allows passengers
approximately an hour-and-a-half on Wednesdays or three
hours on Saturdays to visit the site. Members of the Mowa-
chaht/ Muchalaht First Nation give excellent guided tours.

that accompanied their initial encounter. On March 29,
1778, having been battered for days by brutal weather
during his dash north to the Bering Strait, celebrated
English adventurer and surveyor James Cook took
advantage of moderating winds to seek sanctuary in
a deep sound midway up Vancouver Island.

Intending to guide their visitors into the harbor,
the Mowachaht, "paddl[ing] in most excellent time,
the foremost man every 3rd or 4th Stroke making flour-
ishes with his paddle", according to the diaries of 2nd
Lieutenant James Burney, sped out in their magnificent canoes to meet Cook's ships, the HMS *Resolution*
and HMS *Discovery*. Circling them as they sailed on
an evening breeze toward the tip of Bligh Island, the
paddlers called *"itchme nutka, itchme nutka"*, meaning
"go around" to the sheltered waters at Yuquot. Cook
misunderstood, believing "Nootka" to be the name
of the place. And Nootka Sound it remains today.

That meeting, which lasted only until the weather
improved in early April, initiated many firsts, including
the first sustained contact with Europeans by the
original inhabitants of B.C. The key word here is
"sustained", for the Mowachaht had twice—in 1774
and 1775—traded with Spanish ships off the southwest shore of Nootka Island, and may have been
aware of the Russians who, led by a Dane, Vitus
Jonassen Bering, had reached Alaska in 1741.

Yuquot was also the site of the first small European settlement, created by the Spanish under Juan
Francisco de le Bodega y Quadra between 1789 and 1795.
And during these early years, it was the main trading
centre on the West Coast. For these reasons, though
fewer than twenty members of the Mowachaht/
Muchalaht First Nation now live there, Yuquot has
long been recognized as a national historic site.

Whales of the Pacific

THE NUMBERS are not what they once were, but the western Pacific populations of gray and humpback whales are definitely recovering from the unimaginable slaughter that took place off the west coast of North America.

A century ago, the gray whales of the Pacific Ocean were all but annihilated with a hunt that pursued them even in their calving lagoons off Mexico's Baja California peninsula. (The Atlantic gray whale population had been hunted to extinction, mainly by Basque hunters, a century earlier.) Finally, with their numbers reduced to just a few hundred animals, the International Whaling Commission brought a halt to the Pacific harvest in 1947. Since then, the eastern Pacific population has slowly recovered to an estimated 19,000 to 23,000 animals today. The western Pacific population, in the waters off Korea, however, numbers no more than 200 animals and is greatly imperilled.

Whales in the eastern Pacific begin life in January, in the warm lagoons of Baja California. The four-or five-metre-long calves weigh a tonne at birth, and gain twenty-five kilos a day on their mother's rich milk. In mid-February, the main herds begin the long migration to their feeding grounds in the Bering Sea and Arctic Ocean. The calves and their mothers stay for another six weeks before following them north. Hugging the coastline, swimming in small groups, they cover between sixty and eighty kilometres a day and by mid-March, can be seen off the coast of British Columbia, spouting, fluking—displaying their huge tails as they dives to feed—or breaching. Most continue on, arriving in the Arctic waters, where they will feed during the long summer days, in June. But some, a resident population of perhaps fifty, stay off the coast of Vancouver Island all summer. Feeding on small shrimp-like crustaceans that swarm near kelp beds and reefs, or worms that burrow in the sea floor mud, gray whales somehow eat enough to reach a length of thirteen metres and a weight of more than thirty tonnes. Females, slightly larger than the males, begin breeding when they are about eight and give birth to a calf every two years. Unless they are tangled in commercial fish nets and drowned, or killed in collisions with a ship, both sexes live to be about sixty.

Humpback whales are about the same length and weight, but have stout bodies and long curved flippers; by comparison, gray whales are slender and graceful. Surprisingly, perhaps, humpbacks seem more athletic, often somersaulting completely out of the water or leaping with flapping flippers, as though trying to fly.

Granted international protection in 1965, humpbacks, whose magnificent songs ring through the warm ocean waters where they breed, have been slower to recover their numbers. Today, the north Pacific population is estimated at between 6,000 and 8,000, according to Environment Canada, with as many as 2,000 off the coast of British Columbia. Humpbacks are also found off the waters of Hawaii and in the Atlantic and worldwide,

Dennis Fast

their numbers appear to be growing. Remarkably, though the males in each population sing the same songs, different populations sing different songs, with each song slowly changing over time. The singing occurs in warm or temperate waters; in B.C.'s cold waters, the whales make rougher scraping and groaning sounds, perhaps to locate food.

When feeding, humpback whales sometimes cooperate; encircling a school of herring or mackerel and blowing bubbles to create a "net" around the fish. They then swim through the clustered school with their mouths open. Like gray whales, they also feed on krill and crustaceans.

For millennia, gray whales or *mah-ac* were of enormous importance to the Nuu-chah-nulth people. Not only did they provide crucial meat, oil, fat and bone, these magnificent mammals were also held in great reverence. Hunting them, therefore, was done with much ritual and ceremony.

The hunt was preceded by ritual cleansing, sometimes so arduous that it caused the hunters' skin to bleed. And prayers were said, including some that were almost sexual in nature. One such prayer, quoted in *Canada: A People's History*, went in part, "May I cause the whale to emerge from the head area, may it come up my canal. May it listen to my words … may it surface right where I am on the water, that it may wait for me. May I cause it to become lame when I start paddling after it. May I bend its mind to my advantage. That the females would want me, that the female whales would want to marry me …"

The Nuu-chah-nulth canoes, which seemed so large on land, so painstakingly carved and beautifully painted, were eclipsed by the great whales on the open ocean. It therefore took great courage to hunt *mah-ac*; only the most worthy could launch the first harpoon.

When struck, a whale would dive, its plunge halted by air-filled sealskins that were attached to the long harpoon rope. The struggle that followed sometimes took hours, before at last the great animal was towed back to the village; there a successful hunt was celebrated with great ceremony and prayers of thanks. But despite their prowess, training and preparations, the Nuu-chah-nulth were not always successful; those who failed gave their lives to the sea.

Very few cultures even contemplated hunting the great whales on the open Pacific. Those that did went about the task with both preparation and prayers.

Glossary

adze: An axe-like tool for trimming and smoothing wood

alluvial: Found in or made up of alluvium, the sand or silt that is gradually deposited by moving water, such as a river, or at the margins of a lake

ammonoids: Extinct mollusks of the subclass Ammonoidea; see ammonite

ammonite: The flat, coiled, chambered shell of any of a variety of extinct squid-like creatures found as a fossil in Mesozoic formations

amphipod: Crustaceans having a vertically thin body, with one set of legs for walking and jumping, and another set of legs for swimming

anadramous: Referring to fish that spend part of their lives in the ocean, but are born and spawn in fresh water

anthropomorphism: Attributing human characteristics to non-human or inanimate objects

anoxic zone: A region in which oxygen is absent

aqueduct: A passage or canal in which water is carried a considerable distance

Archaean: Of or belonging to the earlier of the two divisions of Precambrian time, from approximately 4 to 2.5 billion years ago, marked by an atmosphere with little free oxygen, the formation of the first rocks and oceans, and the development of unicellular life. Also of or relating to the oldest known rocks, which are predominantly igneous in composition

argillite: A hardened mudstone, midway between shale and slate, which possesses no truly slate-like cleavage (a tendency to split along crystalline planes, yielding a smooth surface)

arthropod: Invertebrate animals having segmented or jointed bodies and hollow, jointed legs

asl: Above sea level

atlatl: (Aztec) " Athrowing stick" or spear thrower; comprised of wood or bone, it was used to maximize the distance of a tipped with a medium-sized stone point on a detachable shaft

atoll: A roughly circular or horseshoe-shaped coral reef enclosing a shallow lagoon, commonly constructed on the tops of individual volcanoes of volcanic island arcs

avens: A plant of the rose family, with compound leaves and variously colored, usually yellow flowers

batholith: A large body of intrusive rock

Beringia: The Bering Strait land bridge connecting Asia and North America, which appears when sea levels drop significantly during periods of glaciation

bifacial tools: Stone tools, including spearpoints and knives as well as later arrowheads, that have been worked on both sides

BP—Before Present: A system of archaeological or geological dating in which the age of artifacts, cultures or events are given in years prior to the present; the year 1950 was chosen to mark the present

brachiopods: A group of nearly extinct bivalve organisms that reached its peak in numbers and diversity during the Paleozoic era

braided river: A river that flows in several dividing and reuniting channels resembling the strands of a braid. Such streams occur where more sediment is brought to any part of a stream than it can remove. The building of bars of sediment becomes excessive, forcing the stream to develop an intricate network of interlacing channels

calcite: Calcium carbonate, the mineral that forms limestone

caisson: A watertight box that can be pumped dry to allow excavation of near-shore seabeds or lake beds.

Cambrian: (542—490 million years ago) The first geologic period in the Paleozoic era, marked by the development of a profusion of marine animals, especially trilobites and brachiopods

caprock: A relatively impervious rock layer immediately overlying a deposit of softer rock or minerals such as oil, gas or salt

capybara: A short-tailed, partially web-footed semi-aquatic South American rodent, found in and around tropical wetlands

carbon dating: The method used to establish the approximate age of fossil remains by measuring the amount of radioactive carbon 14 remaining in the object. The atoms of carbon 14 decay, losing an electron (to become ordinary nitrogen or nitrogen 14) over a predetermined length of time. Carbon 14 has a half-life of about 5,730 years, meaning it takes that long for half the atoms in any quantity of carbon 14 to decay. In other words, after 5,730 years the remains of an organism will contain only one-half the carbon 14 the living organism did, and after 11,460 years only one-quarter as much, and so on. Thus, up to a point, (generally material less than 40,000 years old), organic substances can be reasonably accurately dated by measuring the remaining carbon 14.

carbonate: Carbonates are sedimentary rocks formed by the organic precipitation from marine waters of mineral carbonates of calcium and magnesium

cataract: A large waterfall, or any strong rush of water

caldera: A more or less circular volcanic depression in an area of volcanic activity, with a diameter many times larger than the volcanic vent

Cenozoic: (66.5 million years ago to present) The latest era of geological time, following the Mesozoic era . It is characterized by the development of many types of mammals

cephalopods: Marine mollusks with distinct heads, highly-developed eyes and varying numbers of arms with suckers, like an octopus or squid

chalcedony: A kind of quartz with the luster of wax. Though often grayish or milky, it can be variously colored.

chert: A dense sedimentary rock consisting of extremely fine, interlocking crystals of quartz, usually occurring as nodules or beds within carbonates like limestone

chinook: The warm, dry wind that blows intermittently down the east side of the Rockies during winter and early spring

cirque: A steep-walled, amphitheatre-like recess occurring at high elevations on the side of a mountain, commonly at the head of a glacial valley. It is formed from a mountain glacier's erosive carving.

climatology: The science of climate and climatic phenomena

Columbian mammoth: One of the largest mammoths to inhabit North America, it stood about four metres at the shoulder. Unlike the smaller woolly mammoth, it was virtually hairless, much like today's elephants, and could survive only in climates that were relatively benevolent.

conglomerate: A coarse-grained rock of rounded fragments in a finer matrix

Cordilleran ice sheet—see Wisconsin glaciation

Cretaceous: (145—66.5 million years ago) The third and last period of the Mesozoic era, it was marked by the final flowering of the dinosaurs, the development of early mammals and flowering plants, the deposit of chalk beds and the decline and demise of toothed birds and ammonites

crinoids: A class of marine animals with movable arms, often attached to the sea-floor by a long stalk. Also called "sea lilies", they evolved during the Cambrian period

culturally modified trees (CMTs): Living trees from which sections of bark or wood have been harvested in a way that allows the tree to heal itself. Also, living trees that have been modified in some way that benefits humans

cyanobacteria: A class of bacteria which use oxygen to produce photosynthesis. Commonly called blue-green algae

Dendrochronology: The science of dating events or climate change variations by studying growth rings in trees and aged wood

Devonian: (408—360 million years ago) The middle period of the Paleozoic era, which was marked by an abundance of fishes and the appearance of the first authentic land plants and amphibians. So named because its rocks were first studied in Devonshire

dolomite: Rock consisting of calcium and magnesium carbonate

earth lodge: See kekuli

ecosystem: A system made up of a community of animals, plants and bacteria and its interrelated physical and chemical environments

Ediacaran period: (600—542 million years ago) Earlier known as the Vendian period, it was recently internationally recognized as the final period in the Proterozoic era, immediately preceding the Cambrian.

elasmosaur: A large plesiosaur, with a long neck and small head, which lived and hunted in the ocean

Eocene: (55.5—33.7 million years ago) The second epoch in the Tertiary period of the Cenozoic age. It was marked by significant global warming, which peaked about 50 million years ago (the Eocene Climatic Optimum), and an explosion of plant and animal species that can be traced directly to many of those that inhabit the Earth today. Eocene comes from the Greek eos, meaning "dawn", and ceno, meaning "new"

eon: The longest division of geologic time, containing two or more eras; also, an indefinitely long period of time

erratic: A large boulder that has been carried to its resting place by glacial ice. Erratics that have come a great distance can usually be recognized by their difference in composition from the local bedrock

escarpment: A steep slope or cliff formed by erosion or by uplift or faulting

eulachon: Small anadromous fish that feeds on plankton, fish eggs and insect larvae. Its high oil content made it an important fish for native British Columbians, who harvested it in large numbers as it spawned. Also called "candlefish", since dried fish can be burned like a candle

eutrophic: A body of water, especially a lake or pond, rich in nutrients that cause excessive growth of aquatic plants, especially algae. The resulting bacteria consume nearly all the oxygen

fjord: A narrow inlet from the sea bordered by steep highlands, believed to have been carved by ancient glaciers. The word is Old Norse, and akin to the Scottish "firth"

flint: A fine-grained, very hard siliceous rock, usually grey, which produces sparks when struck with steel and breaks into pieces with sharp cutting edges

fluvial erosion: Erosion caused by the action of flowing water

floodplain: A plain formed along a river, from sediment deposited by floods

fossils: Any hardened remains or imprints of plant or animal life from a previous geologic period, preserved in the earth's crust

freshet: A sudden rush of water down a river or stream resulting from heavy rain or the spring thaw

gastroliths: Small, round stones swallowed by some animals to aid with digestion of tough plants; also called "gizzard rocks"

geology: The scientific study of the origin, history, structure and processes of Earth, including the structure and development of its crust, the composition of its interior, individual rock types and fossil forms

glacial: Anything produced by a glacier or a glacial period

glacial till: Glacial matter composed of clay, stones, gravel, boulders, etc.

glaciation: Specifically, the formation, movement and recession of ice sheets, but also referring to the geological processes associated with glacial activity, including erosion and deposition, and the resulting effects of such action on Earth's surface

glacier: A large mass of long-lasting ice that forms on land by the recrystallization and compaction of snow and moves slowly downslope or outward in all directions because of its own weight

gorge: A deep narrow pass between steep heights

gneiss: A layered, coarse-grained rock rich in quartz and feldspar

grasslands: A region where grass predominates; a prairie

Gulf of Georgia phase: The latest of the sophisticated Lower Mainland cultures, beginning about 1,200 BP and continuing until contact. Very similar to the preceding Marpole phase, it is marked mainly by changes in burial practices, differences in style, including use of bone and antler weapon points, and bow-and-arrow hunting technology

Gulf Stream: A broad "river" or current of warm water that flows in a counterclockwise motion from the Gulf of Mexico to Northern Europe, significantly warming countries such as France and Britain. Also called the sub-polar gyre

half-graben: A subsided block of land, or downward moving fault resulting from fractures or crustal stretching

hanging valley: The result of glaciation, where an upper valley abruptly ends at the edge of a deeper, wider valley below.

Holocene epoch: The present epoch of the Quaternary period (our glacial age), extending from the close of the Pleistocene epoch about 10,000 years ago. The realization that we are still in the midst of a glacial age may make the Holocene epoch eventually redundant

hoodoo: A pillar of stone created when a hard caprock protects soft underlying rock from erosion. The word was originally associated with voodoo culture, and meant "bad luck". Because hoodoos sometimes occur in large numbers in eroded valleys and canyons, to European minds the resulting forest of otherworldly formations seemed sinister and magical

Hypsithermal interval: The warmest period in our current interglacial, between about 8,000 and 5,000 BP. Global temperatures were about 1.5EC warmer than they are today. It was followed by the Neoglaciation, which continues—with warmer intervals—today (see graph on page 20)

Ichthyosaurs: These giant sea-going reptiles flourished through the Jurassic period and into the Cretaceous. They were torpedo-shaped with streamlined bodies and stabilizing dorsal fins on the back. Though popularly associated with dinosaurs, they were not dinosaurs and were not closely related to dinosaurs. Evidence suggests ichthyosaurs gave live birth, which dinosaurs did not do.

Imperial mammoth: The largest North American mammoth, standing more than four metres at the shoulder. Like the Columbian mammoth, the imperial mammoth was likely descended from the Eurasian steppe mammoth. In fact, there are those who believe imperial and Columbian mammoths are the same species.

Intermontane Belt: A region of volcanic plateaus and highlands, often collectively called the Interior Plateau, that lies between the Coast Mountains on the west and the Columbia Mountains on the east and runs from south-central Yukon to northern Washington State

Insular Superterrane: The collective terranes (including Wrangellia, the main part of Vancouver Island) that were welded onto the edge of B.C. about 90 million years ago. This superterrane stretches from the eastern edge of the Strait of Georgia to the subduction zone west of Vancouver Island, and runs north to southeastern Alaska. It includes the Gulf Islands and Haida Gwaii.

interglacial: The relatively warm, ice-free interval between two glaciations, or the results of that interval. Our current Holocene epoch is an interglacial interval

intermontane: A region between or among mountains

ironstone: Any rock rich in iron

island arcs: A belt of volcanic islands created by an undersea vent. Moving with the ocean plate, the island arc eventually meets a continental plate. The volcanic islands are peeled off and sit at the edge of the continent above a subduction zone. Japan and Alaska's Aleutian islands are examples. The belt of islands is usually curved, with the concave side facing the continent

isostatic depression (also isostatic rebound): The depression of the Earth's crust caused by the weight of an ice sheet; depression is generally about one-third of the thickness of the ice. (Following the melting of the ice, the crust slowly returns to its original position.)

jökulhaup: (Icelandic) A sudden outburst of water released by a glacier, generally by the failure of an ice dam

Jurassic: (213—145 million years ago) The second of period of the Mesozoic or dinosaur era, following the Triassic and preceding the Cretaceous, it lasted about 60 million years (between 208 and 140 million years ago) and was characterized by the dominance of dinosaurs and the appearance of flying reptiles and birds.

kame: A hill or short, steep ridge of stratified sand or gravel deposited by glacial ice

karst: Topography, characterized by sinkholes, caves and underground streams formed by the dissolution of rock in water. Most topography of this type develops in regions of limestone

kekuli: A circular or oval pithouse, with a foundation that was dug into the ground. This type of housing was used throughout the southern Interior Plateau, beginning almost 4,000 years ago

kettle: A steep-sided, usually basin- or bowl-shaped depression in glacial drift deposits, often containing a lake and formed by the melting of a large, detached block of ice left behind by a retreating glacier

kokanee: Freshwater salmon; the word is Ktunaxa for "red fish"

labret: A small ornament of carved stone worn through the lip or nose; similar to today's nose or lip rings

Laurentide ice sheet: See Wisconsian glaciation

leister spear: A spear with three or more prongs, used for spearing fish

limestone: Rock consisting mainly of calcium carbonate

lithics (or lithic materials): Objects made of stone

Little Ice Age: A period of global cooling between 650 and 100 years ago (or 1350 and 1900 AD), which followed almost immediately on the heels of the Medieval Warm Period

Locarno Beach phase: Named for a site on English Bay, it is the earliest of the sophisticated Lower Mainland cultures (3,500—2,500 BP), which were marked by the development of a socially ranked society. See also Marpole and Gulf of Georgia cultures

loess: A loose deposit of silt and clay deposited by wind

mammoths: Later arrivals on the Cenozoic scene than mastodons, mammoths bear a close relationship to modern-day elephants and were, in various lineages, commonplace on the planet. The woolly mammoth (Mammuthus primigenius), with its high domed head, long sloping back and body covered with dense black hair, was profuse in the tundra belts around glaciated North America but, like mastodons, may have been hunted to extinction by humans. Recent discoveries indicate that mammoths continued to live on remote islands in the Canadian Arctic until almost 4,000 BP. Canadian scientists have been mapping the mammoth genome, raising the question of whether it might be possible to recreate an extinct mammal.

Marpole phase: Named for a site on the north shore of the Fraser Delta, the Marpole phase (2,400—1,200 BP) of the sophisticated Lower Mainland culture was characterized by the widespread appearance of large houses, standardized art forms, and elaborate burials

mastodon (or mastodont): The dominant megafauna of the Miocene epoch, 24 million to five million years ago, mastodons endured in North America until about 10,000 years ago, surviving the last glaciation, browsing in herds in spruce woodlands. As high as three metres (10 feet) at the shoulder, the elephant-resembling mastodons were covered in a long coat of shaggy hair, had generally long heads held low, and were possessed of massive inward-curving tusks. Mastodons may have been hunted to extinction by early man.

Medieval Warm Period: A period of global warming between 1,200 and 700 years ago (or 800 and 1300 AD) marked by significant warming of the Northern Hemisphere

megafauna: A term used to describe large mammalian life-forms that once dominated the North American continent but are now extinct

Mesozoic: (245—65 million years ago, the "Age of Medieval Life") The geologic era after the Paleozoic and before the Cenozoic, a period between 245 million and 65 million years ago

microblade: A long, narrow, sharp-edged piece of stone, generally between three and six millimetres (one-tenth to one-fifth of an inch) wide and from 10 to 30 millimetres (one-half to one-and-a-half inches) long, which was set into bone, antler or wooden tools to use as a cutting edge

midden: An ancient refuse heap

Miocene: (23.8"—5.3 million years ago) The fourth epoch of the Tertiary period in the Cenozoic era, marked by the evolution of many mammals of relatively modern form. From the Greek meion, meaning "less" and ceno, "new"

Mississippian: (360—325 million years ago) The first portion of the coal-forming or Carboniferous period of the Paleozoic era in North America, lasting for about 40 million years

molten: Melted or liquefied by heat

moraine: A mass of rocks, gravel, sand and clay carried and deposited by a glacier along its side, at its lower end or beneath the ice

mosasaurs: Enormous sea-going lizards with a long powerful tails and paddle-shaped legs for swimming, found in Alberta's shallow seas during the Cretaceous period. Like ichthyosaurs, they were not dinosaurs. No fully aquatic dinosaur species is known

muskox: A hardy mammal of Arctic North America and Greenland with a long, coarse, hairy coat, large curved horns and a musky odor. Two subspecies of muskoxen lived in Canada; one, the helmeted muskox, died out in the waning millennia of the last glaciation

nautiloids: A group of cephalopod molluscs with an external, chambered shell that is either straight or coiled. Though very diverse in the past, only the pearly nautilus, survives today.

nephrite: A white to dark green variety of jade

nunatak: An isolated knob or peak of land that projects prominently above the surface of a surrounding glacier

obsidian: Volcanic glass, often dark-colored, but occasionally found in lighter colors as well

ochre: An earthy clay colored by iron oxide, usually yellow or reddish brown, used for paint by peoples around the world

Ordovician period: (490—443 million years ago) The second earliest period of the Paleozoic era, it follows the Cambrian. Named after a Celtic tribe called the Ordovices

Oligocene: (33.7—23.8 million years ago) The third epoch of the Tertiary period in the Cenozoic era

oligotrophic: Poor in nutrients and plant life

outwash: Sand and gravel deposited by meltwater streams at the edge of glacial ice

Paleolithic: Of an Old World cultural period, before the Mesolithic, characterized by the use of flint, stone and bone tools, hunting, fishing and gathering of plant foods; also called the Stone Age

paleomagnetism: The alignment of iron and nickel grains in rock with the earth's magnetic poles, fixed at the time of that rock's formation. Paleomagnetic dating studies the residual magnetism in rocks to try to determine their age or the time of their creation paleontology (also palaeontology): The study of fossils and ancient life forms

Paleozoic: (542—248 million years ago) The geologic era between the Precambrian and Mesozoic eras, characterized by the development of the first fish, amphibians, reptiles and land plants

Pangea (also Pangaea): The name given to the giant supercontinent (made up of all the world's continents) that had assembled by about 300 million years ago and began to break up about 200 million years ago. The initial breakup resulted in the separation of Laurasia, including North America and Eurasia, from Gondwanaland, which contained South America, Africa, India, Australia and Antarctica.

petroglyph: A symbol or picture carved into rock

pictograph: A picture or symbol painted onto rock using ochre or other pigments and a variety of techniques

phylogeny: The evolutionary development of a species or genus plate tectonics: The dynamics of the movement of the Earth's crustal plates

Pleistocene: (1.9 million to 10,000 years ago) The first epoch of the current Quaternary period in the Cenozoic era, characterized by the periodic incursion and recession of continental ice sheets and by the appearance of modern humans

plesiosaurs: Giant marine reptiles with large flippers instead of limbs. Like penguins, plesiosaurs would skim across the world's ancient seas in their search for food using their flippers like paddles

plutons: Igneous intrusions into granite, formed when the rock was beneath the Earth's surface

Precambrian: The stage of Earth's history between its formation some 4.6 billion years ago, and about 542 million years ago when life began to dramatically increase in both numbers and diversity.

The Precambrian era generally divided into two eons, the Archean, ending about 2.5 billion years ago, and the Proterozoic, ending 542 million years ago.

primordial: First in time, existing from the beginning

proglacial lakes: Lakes produced by melting glaciers, often in front of the glacier

pterosaurs: Close relatives of the dinosaurs, pterosaurs were the first vertebrates to take to the air, gliding on wings of skin stretched along the length of the greatly elongated fourth finger of each hand, rejoining the body at thigh level. Pterosaurs evolved 70 million years before archaeopteryx, the true flying dinosaur, and had wingspans as long as 15 metres

Quaternary: The second period of the current Cenozoic era, from about 1.8 million years ago to the present. It contains two epochs: the Pleistocene and the Holocene

quartz: A mineral compound of silica and oxygen, most commonly colorless or white, originally formed during crystallization of molten magma and a common constituent of igneous, metamorphic and sedimentary rock

quartzite: A hard, unmetamorphosed sandstone, consisting chiefly of quartz grains cemented together by crystalline quartz or a dense conglomerate rock comprised of sand, silt and pebbles compressed and cemented by time and pressure

Quesnellia: The Quesnel terrane, one of a series of small volcanic terranes of late Devonian or early Jurassic rock that were accreted to the western edge of the old North American continent about 180 million years ago. They are found in the Interior Plateau.

radiocarbon dating: See carbon dating

recessional moraines: An end or lateral moraine constructed during a significant pause in the retreat of a glacier

receptaculites (Fisherites recptaculitids): An extinct form of algae that lived in large globular colonies about 450 million years ago. Also called "sunflower coral"

riparian: Adjacent to, or living on, the bank of a river, lake or pond

ripple marks: Parallel, small-scale ridges and hollows produced by currents acting on sand or silt, such as those seen on modern beaches

Rodinia: A supercontinent that is believed to have formed about 1.1 billion years ago and to have broken up into eight continents around 700 million years ago.

sandstone: A sedimentary rock composed of rounded to angular fragments of sand, set within a matrix of silt clay, and cemented by calcite, silica or iron oxides. The sand particles most commonly consist of quartz

ist: A metamorphic crystalline rock that can be split along oximately parallel lines

epemc: The traditional name of the people of the Thompson r area; also called the Shuswap

ment: Matter deposited by wind or water

entine: 1) A rock, usually green, yellow or brown mottled h red, composed primarily of hydrated magnesium silicate. 2) description of the sinuous course of a river or a path

shale: A fine-grained sedimentary rock formed from the consolidation of clay and silt

silt: Sediment suspended in stagnant water or carried by moving water that often accumulates on the bottom of rivers, lakes or bays smolts: Salmon young

spelunker: One who explores and studies caves

steatite: Easily carved soapstone; also, when ground, used for thickening paints

stromatolite: A laminated sedimentary rock structure formed primarily during the Proterozoic eon in shallow pools by mats of cyanobacteria or blue-green algae. The algae trapped layers of silt, especially calcium carbonate. These wavy or round formations provide evidence for dating the first life forms on Earth

stromatoporoids: Extinct marine, reef-building organisms of uncertain biological affinity. They secreted skeletons of calcium carbonate in a wide variety of shapes and were especially abundant from Ordovician through Devonian time, from 505 million to 360 million years ago

subduction zone: The area where the ocean plate meets and begins to slide under the continental plate

subtropical: A region bordering a tropical zone

tectonic: Pertaining to, causing or resulting from the structural deformation of Earth's crust

topography: A detailed description or drawing of the surface features of a place or region

travois: A sledge, pulled by a dog or, later, a horse, created by slinging netting or a platform between two long poles

Triassic: (248—213 million years ago) The earliest part of the Mesozoic era. It was during this period that the supercontinent Pangea came apart, giving birth to our modern continent of North America.

trilobite: A segmented, three-lobed, bottom-dwelling marine arthropod that lived during the Paleozoic era, most abundantly during the Cambrian and Ordovician periods

tsunami: A huge ocean wave or wall of water created by an earthquake or volcanic eruption. These waves may reach enormous heights and cross entire oceans

tufa: Porous, spongy calcium carbonate deposited around the outlet of a deep mineral spring

ungulate: Having hoofs, or belonging to a group of mammals having hoofs

unifacial tools: Stone tools which have been worked on only one side; these are usually older than bifacial tools and an indication of a very early Paleolithic or Stone Age society

varves: An annual sequence of two deposits of sediment in a glacial lake. During the summer, when active melting takes place, sediments are generally thicker and coarser. In the winter, the lake freezes, allowing finer sediments to settle to the bottom. The two layers taken together allow glaciologists to determine the duration of a glacial lake

vertebrates: Animals, including reptiles, fishes and birds, that have a bony or cartilaginous spinal column

viviparous: Giving birth to live young

volcanic terranes: (see island arcs)

Wisconsin glaciation (or Wisconsinan): The last and one of the largest incursions of ice during our current "ice age". This latest glacial age (the Quaternary period) began about 1.9 million years ago. Since then, great sheets of ice have advanced and retreated at least 17 and perhaps as many as 20 times. The Wisconsin glaciation seems to have had two distinct phases: the Early Wisconsinan, from about 75,000 BP to about 65,000 BP, and the Late Wisconsinan, which began about 25,000 years ago and reached its estimated maximum about 18,000 years ago. Because of its geographical location west of the Rockies, British Columbia was invaded by the Cordilleran or mountain ice sheet, while the rest of Canada was covered by the much larger Laurentide or continental ice sheet, which had its genesis over Hudson Bay. The Cordilleran ice sheet eventually covered all but the highest mountains from the Rockies west to the Pacific coast. However, recent studies have shown that parts of British Columbia, including southern Vancouver Island and parts of the Thompson Okanagan region, were not covered by ice until very late, perhaps 15,000 BP. By 13,000 BP, the ice was beginning its retreat and in the southern reaches of the province, the Pleistocene epoch (the ice age part of the Quaternary period) was over by about 11,000 years ago, giving way to the Holocene epoch (or interglacial period) in which we live.

woolly mammoth: Covered with a coat of thick dun-colored fur, the woolly mammoth—the best-known of all the mammoths—was smaller than the Columbian mammoth. It evolved in Siberia, but eventually thrived across the Northern Hemisphere from Ireland to North America's arctic regions.

zircon dating: Zircon is a pervasive mineral that has become important for radiometric dating. Containing amounts of uranium and thorium (from 10 ppm up to 1% wt), it can be dated using modern analytical techniques. Since zircons have the capability to survive geologic processes like erosion, transport and even metamorphism, they are used to date some of the world's oldest rocks. The oldest found so far are from Western Australia, with an age of 4.404 billion years, interpreted to be the age of crystallization. Their oxygen isotopic composition also seems to indicate there was water on the Earth's surface at that time.

Bibliography

A Geoscience Guide to the Burgess Shale, by Murray Coppold and Wayne Powell, The Yoho-Burgess Shale Foundation, Field, BC. 2000

A Traveller's Guide to Aboriginal B.C., by Cheryl Coull, Whitecap Books, Vancouver, and Beautiful British Columbia Magazine, 1996

After the Ice Age: The Return of Life to Glaciated North America, by E.C. Pielou, The University of Chicago Press, 1991

Alberni Prehistory: Archaeological and Ethnographic Investigations on Western Vancouver Island, by Alan D. McMillan and Denis E. St. Claire, Alberni Valley Museum and Theytus Books, Penticton, BC. 1982

Artifacts of the Northwest Coast Indians, by Hilary Stewart, Hancock House Publishers, Saanichton, BC. 1973

Current Research Reports, Roy Carlson (ed.), Simon Fraser University, Burnaby, BC. 1976

Geology of British Columbia: A Journey Through Time, By Sydney Cannings and Richard Cannings, Greystone Books, Douglas & McIntyre Publishing Group, Vancouver. 1999

Early Human Occupation in British Columbia, Roy Carlson and Luke Dalla Bona (eds), UBC Press, Vancouver, 1996

Gabriola: Petroglyph Island, by Mary and Ted Bentley, Sono Nis Press, Victoria. 1981

Handbook of the Canadian Rockies, by Ben Gadd, Corax Press, Jasper, AB. 1995

Hot Springs of Western Canada: A complete guide (2nd ed), by Glenn Woodsworth, Gordon Soules Book Publishers Ltd. West Vancouver. 1999

In Search of Ancient Alberta, by Barbara Huck and Doug Whiteway, Heartland Publications, Winnipeg, MB.1998

Late Prehistoric Cultural Horizons on the Canadian Plateau, by Thomas H. Richards and Michael K. Rousseu, Simon Fraser University, Burnaby, BC. 1987

Life in Stone: A Natural History of British Columbia's Fossils, Rolf Ludvigsen (ed), UBC Press, Vancouver, 1996.

Plant Technology of First Peoples of British Columbia, by Nancy J. Turner, UBC Press, 1998.

Plants of Coastal British Columbia (revised edition), Jim Pojar and Andy MacKinnon (eds.), B.C. Ministry of Forests and Lone Pine Publishing, Vancouver. 2004

Plants of Southern Interior British Columbia and the Inland Northwest, Roberta Parish, Ray Coupé and Dennis Lloyd (eds.), B.C. Ministry of Forests and Lone Pine Publishing, Vancouver. 1996

Roadside Geology of Southern British Columbia, by Bill Mathews and Jim Monger, Mountain Press Publishing Compan Missoula, MT. 2005

Stone, Bone, Antler and Shell: Artifacts of the Northwest Coast, by Hilary Stewart, Douglas & McIntyre Ltd., Vancouver, 1996

The Ancient Past of Keatley Creek, Vols I and II, Brian Hayden (ed.), Archaeology Press, Simon Fraser University, Burnab BC 2000

The Archaeology of Kamloops, by Robert L. Wilson and Catherine Carlson, Simon Fraser University, Burnaby, BC. 1980

The Belcarra Park Site, by Arthur S. Charlton, Simon Fraser University, Burnaby, BC. 1980

The Prehistoric Use of Nephrite on the British Columbia Plateau, by John Darwent, Archaeology Press, Simon Fraser University, Burnaby, BC. 1998

The Sooke Story: The History and the Heartbeat, by Elida Peers, Sooke History Book Committee, Sooke, BC. 1999

Vancouver, City on the Edge, by John Clague and Bob Turner, Tricouni Press, Vancouver, 2003

West Coast Fossils: A Guide to the Ancient Life of Vancouver Island (2nd. ed.), By Rolf Ludvigsen & Graham Beard, Harbour Publishing, Madiera Park, BC. 1997.

Also highly recommended is the *Backroad Mapbook* series, by Russell and Wesley Mussio. Mussio Ventures has now published detailed and accurate mapbooks of most of the regions of British Columbia and is now working on Alberta. And *Shale: Journal of the Gabriola Historical and Museum Society*, has many interesting articles, even if you're not an islander.